ALASKA'S FOREST RESOURCES

Volume 12, Number 2 / 1985
ALASKA GEOGRAPHIC®

To teach many more to better know and use our natural resources

Editor: Penny Rennick
Associate Editor: Kathy Doogan
Designer: Sandra Harner
Maps and charts by: Jon.Hersh

ALASKA GEOGRAPHIC®, ISSN 0361-1353, is published quarterly by The Alaska Geographic Society, Anchorage, Alaska 99509-6057. Second-class postage paid in Edmonds, Washington 98020-3588. Printed in U.S.A.

Registered trademark; Alaska Geographic, ISSN 0361-1353; Key title Alaska Geographic.

THE ALASKA GEOGRAPHIC SOCIETY is a nonprofit organization exploring new frontiers of knowledge across the lands of the polar rim, learning how other men and other countries live in their Norths, putting the geography book back in the classroom, exploring new methods of teaching and learning — sharing in the excitement of discovery in man's wonderful new world north of 51°16′.

MEMBERS OF THE SOCIETY RECEIVE *Alaska Geographic®*, a quality magazine which devotes each quarterly issue to monographic in-depth coverage of a northern geographic region or resource-oriented subject.

MEMBERSHIP DUES in The Alaska Geographic Society are $30 per year; $34 to non-U.S. addresses. (Eighty percent of each year's dues is for a one-year subscription to *Alaska Geographic®*.) Order from The Alaska Geographic Society, Box 4-EEE, Anchorage, Alaska 99509-6057; (907) 563-5100.

MATERIAL SOUGHT: The editors of *Alaska Geographic®* seek a wide variety of informative material on the lands north of 51°16′ on geographic subjects — anything to do with resources and their uses (with heavy emphasis on quality color photography) — from Alaska, northern Canada, Siberia, Japan — all geographic areas that have a relationship to Alaska in a physical or economic sense. We do not want material done in excessive scientific terminology. A query to the editors is suggested. Payments are made for all material upon publication.

CHANGE OF ADDRESS: The post office does not automatically forward *Alaska Geographic®* when you move. To ensure continous service, notify us six weeks before moving. Send us your new address and zip code (and moving date), your old address and zip code, and if possible send a mailing label from a copy of *Alaska Geographic®*. Send this information to *Alaska Geographic®* Mailing Offices, 130 Second Avenue South, Edmonds, Washington 98020-3588.

MAILING LISTS: We have begun making our members' names and addresses available to carefully screened publications and companies whose products and activities might be of interest to you. If you would prefer not to receive such mailings, please so advise us, and include your mailing label (or your name and address if label is not available).

ABOUT THIS ISSUE:

To compile this issue about Alaska's forest resources, we called upon Walt Matell, a freelance writer and photographer who lives in Ketchikan. Many people from throughout Alaska helped contributing editor Matell with the project, generously offering information and photographs, and reviewing the text. Especially helpful were: Susan Brook, for her contribution to the chapter on growing trees; Kirsten Held of Alaska Loggers Association; Jim Rynearson of Alaska Pulp Corporation; David Wallingford of the Alaska Division of Forestry; Gregory Head of Alaska Timber Corporation; Bill Farr, Al Harris and Ken Winterberger of the Forestry Sciences Laboratory; John Alden, Les Viereck and John Zasada of the Institute of Northern Forestry; Steve Laroe of the Interior Woodcutters Association; Mel Mountain of Louisiana-Pacific Corporation; Ross Soboleff of Sealaska Corporation; Al Pagh of Four Star Lumber Company; Gary Candelaria of Sitka National Historical Park; and Dean Argyle, Carl Holguin, Wini Sidle, Joe Mehrkens, John Raynor and Butch Ruppert of the U.S. Forest Service. Thanks also go to the many contributors who submitted photographs, and to Kathy Lucich and Dale Taylor for reviewing the manuscript.

Editor's note: Terms appearing in **bold face** in the text can be found in the glossary on page 190. The following abbreviations have been used throughout the book: USFS (U.S. Forest Service); BLM (Bureau of Land Management); S&PF (State and Private Forestry). Unless otherwise indicated, all temperatures are given in Fahrenheit.

The Library of Congress has cataloged this serial publication as follows:

Alaska Geographic. v.1-
[Anchorage, Alaska Geographic Society] 1972-
v. ill. (part col.). 23 x 31 cm.
Quarterly.
Official publication of the Alaska Geographic Society.
Key title: Alaska geographic, ISSN 0361-1353.

1. Alaska — Desription and travel — 1959-
—Periodicals. I. Alaska Geographic Society.

F901.A266 917.98′04′505 72-92087
MARC-S
Library of Congress 75[7912]

The cover photo for this issue of *ALASKA GEOGRAPHIC®* is of a slice of Alaska yellow-cedar that is one of several produced for eventual awards to Alaska's Historian of the Year winners. With appropriate salt treatment and curing, plus a plastic finish, the resulting plaque not only represents symbolically a large segment of Alaskan history, it is also of unusual beauty. Alaska yellow-cedar in the fresh state has a delightful aroma and all the clothes protecting characteristics of camphor wood. (Ken Schmidt, staff)

Previous page — **A blend of hardwood and softwood trees creates an autumn mosaic in Denali National Park & Preserve.** (Rollo Pool)

Right — **A Japanese ship takes on logs at Port Chatham. More than 90 percent of the timber exported from Alaska goes to Japan, where the high-quality wood is sought for use in houses, furniture and musical instruments.** (George Ripley)

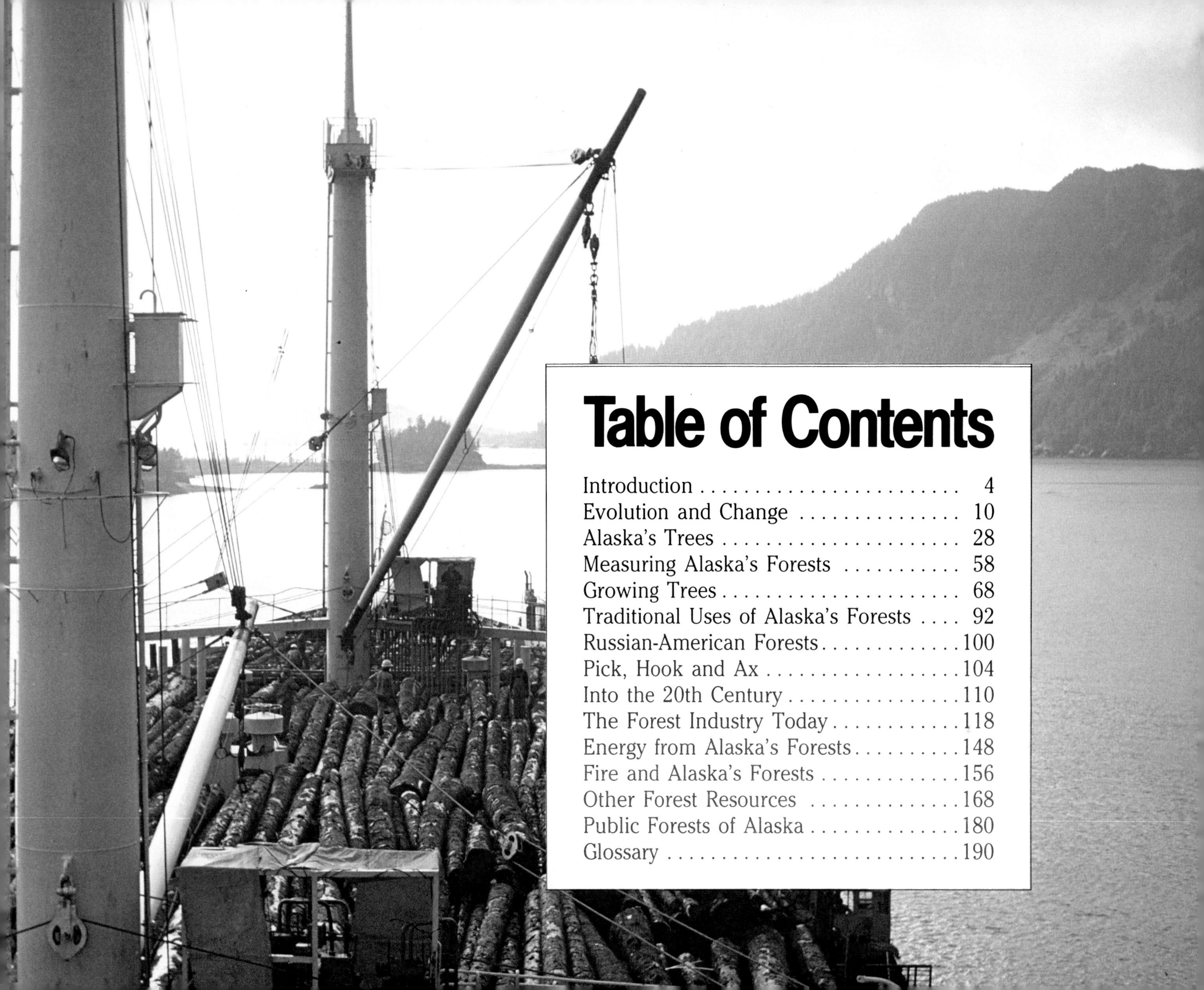

Table of Contents

Introduction

Pacific red elder, also known as elderberry, blossoms in a yard in Homer, on the Kenai Peninsula. Elderberry is common in moist areas along the coast from Ketchikan to the Alaska Peninsula. (Bob & Janet Klein)

Myth number one about Alaska states that ice and snow cover the entire state. Myth number two says that whatever isn't under ice and snow is unbroken forest.

About one-third of Alaska's land area is indeed forested, but the forests are not especially dense except along parts of the coast. Alaska's forests range from open patchwork quilts of hardwoods and conifers on rolling hills in the Interior to luxuriant rain forests on steep mountainsides rising from salt water on the coast. Western hemlock trees no bigger around than a man may be many hundreds of years old on some sites; elsewhere, balsam poplar might reach that size in less than 70 years.

Forests grow slowly by human standards. Mature trees in Alaska today were growing more than 100 years ago when the United States purchased the territory from Russia. When we take measures to improve a forest's yield of timber or other products, we are investing in a resource that probably won't return the investment in our lifetime.

About 3,000 Alaskans are employed in the forest products industry, and many more have subsistence lifestyles or other occupations somehow related to the forests. Salmon fishermen, for example, depend on fish that spawn in forest habitats. Even those whose jobs are not directly related benefit from them daily; try to imagine a naked Alaska — entirely tundra or ice field — without beaver, eagles, shade, or even driftwood!

On the following pages, we present an overview of an Alaskan renewable resource that's often taken for granted, or thought about only as a backdrop for other resources. We cover how forests grow and how people can harvest that growth. We discuss energy, stored in firewood or released in wildfire. A chapter on the 33 tree species native to Alaska gives tips on identification. We hope to give a feeling for how trees work together as biological systems, and how Alaskans' use of forests has changed throughout the years and may change in the future.

Walt Matell
Contributing Editor

An aged cedar tree reaches for the sky above Blank Island in southeastern Alaska. (Don Cornelius)

◄**Wind and salt spray along the coast of southeastern Alaska ensure that this western hemlock will grow slowly, never reaching the heights of its inland cousins.** (Mary Ida Henrikson)

▼**Louisiana Pacific's storage area and dock are prominent in this view of the Ketchikan waterfront taken several years ago. This facility closed in 1983.** (Bruce Katz)

◄**A cloudy sunrise creates an eerie scene over the forest on Gravina Island in southeastern Alaska.** (Tom Buckhoe)

▲**A group of teenagers studies the ecology of a black spruce forest at Nancy Lake Recreation Area, north of Anchorage in the Matanuska-Susitna Valley.** (Nancy Michaelson)

►**Rafts of logs wait in a protected cove to be towed to the nearby pulp mill on Silver Bay, east of Sitka.** (Bruce Katz)

Evolution and Change

by Susan Brook

[**Editor's note:** *Susan Brook is a writer/researcher with a background in forestry and natural resources.*]

The life of most forests appears to be short compared to the . . . time scale of climatic shift.

—J.L. Harper

Climates have indeed shifted in Alaska throughout the ages. During the Quaternary period, the last million years, intervals of generally cooler and drier weather caused at least four major glacial advances. With much of the planet's water tied up in ice during these periods, the sea level lowered enough to expose a land bridge from Alaska to Siberia. Warm intervals between cold cycles melted some of the ice, causing the sea to flood the land bridge.

Ice didn't cover Alaska entirely, even during glacial maximums. A few areas escaped the crush of glaciers: central and western Alaska, the Yukon Delta, parts of the North Slope, mountain peaks protruding above the ice (called nunataks), rocky headlands of some of the islands in southeastern Alaska, and portions of the Gulf Coast. Plants from these **refugia** repopulated terrain laid bare when glaciers receded.

Were there forests in central Alaska during the chilliest parts of the Ice Age, or was the climate there so cold and dry, even in refugia, that only tundra plants could survive? Some researchers have fossil evidence indicating that treeless steppe-tundra existed at this time. They hypothesize that spruce survived only on the Bering land bridge, or south of the ice.

Interstadial stumps, once buried under glacial debris, have been exposed by tidal erosion along the shores of Icy Bay, on the Gulf of Alaska northwest of Yakutat. (Rick Rogers)

Old growth forests, such as this one in southeastern Alaska, contain trees of all ages.
(Don Cornelius)

Others postulate that forests survived in protected valleys of the Interior. Along Alaska's southern coast, trees may have endured the cold in refugia near Prince William Sound or on islands of the Alexander Archipelago, or may have survived only south of the ice in coastal Washington.

The Coastal Forest

Most of the state's commercial forests lie along the somber coastline of southern Alaska. From the Alexander Archipelago to the Kodiak Archipelago, a strip of cold-temperate rain forest extends from mountain slopes to beach fringe, making up the northern extension of the hemlock-spruce forest of coastal Oregon, Washington and British Columbia. Vegetation in Alaska differs because of cooler and wetter conditions.

The climate is maritime, with cold, wet winters and cool, wet summers. The terrain is generally steep; ice fields and jutting peaks separate the rain forests from the drier forests found inland.

In southeastern Alaska, western hemlock and Sitka spruce predominate, with scattered mountain hemlock, western redcedar and Alaska-cedar. Alders abound along streams and beach fringes, and wherever the mineral soil has been disturbed. Huge black cottonwoods grow on floodplains of rivers that drain the coastal mountains. Small quantities of subalpine fir and Pacific silver fir occur at tidewater and near tree line.

Tree Line

The map on page 15 shows the distribution of forests in Alaska today. As one might expect, trees are not found on the Arctic Slope, nor at higher elevations. The general boundary between forested and non-forested areas, **tree line**, does not remain constant, however.

For example, a series of warmer summers the last few decades seems to be causing the tree line to expand on the Seward Peninsula, as well as in the Tanana-Yukon uplands near Fairbanks and in several places in the Alaska Range. In southwestern Alaska, Sitka spruce arrived only 500 years ago, and today the species is pushing the tree line across Kodiak Island at a rate of about a mile per century. At that rate, it may take a long time for the actual tree line to reach its maximum climatic limit. Pacific silver fir appears to be expanding northward in southeastern Alaska for the same reason, dubbed "migrational lag."

Climate probably determines the general positions of tree lines, but scientists disagree about which components of climate are most important. Some believe that tree line is the point where summer temperatures are sufficient to barely keep trees alive; others stress the average temperature of the coldest month; while ecologists working in southwestern Alaska believe that net solar radiation, not air temperature, determines the tree line there.

A solitary paper birch survives beyond tree line on the Porcupine River plateau in northeast Alaska. (Ken Winterberger)

Climate alone does not dictate the forest edge. Availability and mobility of seeds, presence of suitable seedbeds, competitive ability of seedlings to survive on sites already vegetated, and ability of trees to reproduce by non-seed means — such as sprouting from roots or buried branches — all influence the location and advance of the tree line. Human or animal destruction of stands barely able to reproduce may also modify the climatic tree line.

Wildfire, like logging, may permanently displace the tree line if surviving stands are unable to regenerate the burned areas. Fires burning tundra beyond the tree line may, however, actually allow forests to advance by removing competing plants and providing more favorable growing conditions for tree seeds.

The tree line is complicated by the presence of black and white spruce forests along the Kobuk, Yukon and Kuskokwim rivers. How did these isolated fingers of forest develop in these seas of tundra? One possible explanation is that climatic warming allowed seeds transported by wind or the rivers to germinate and establish along the riverbanks. Colder, dryer summers which followed prevented expansion of these floodplain forests. Another theory suggests that ancient forests once covered what is now tundra, and that during a subsequent cooler climate, the forest retreated, leaving these remnants along the moist, protected rivers.

Scrubby shore pine grows in the muskeg near Petersburg. (Walt Matell)

Hemlock seedlings begin their lives on the moss-covered remains of an old tree, appropriately called a "nurse log." (Walt Matell)

West of Prince William Sound, western hemlock drops out of the coastal forest and is replaced by mountain hemlock as the usual companion of Sitka spruce. On Afognak and Kodiak islands, Sitka spruce prevails as the only conifer.

Abundant mosses and bearded lichen festoon branches and fallen logs in the coastal forest. Devil's club is ever present, as are several species of berries, ferns and willows. **Muskeg** bogs, which form where drainage is impeded, pockmark the forest. Peat accumulates under a soggy mattress of mosses, sedges, grasses and heath-type shrubs. Shore pine — coastal lodgepole pine — finds a home on the edges and slightly drier humps in the muskeg.

The Interior Forests

Forests of the Interior, called **taiga**, are links in the chain of evergreen forests that circles the northern hemisphere.

Somewhat similar forests grow in parts of Vermont and Minnesota. Without **permafrost** and extremes of daylight and darkness, however, those forests differ considerably from high-latitude forests.

The coastal forests, the Brooks Range and the coastal tundra along the Bering and Chukchi seas form a boundary for the Alaska

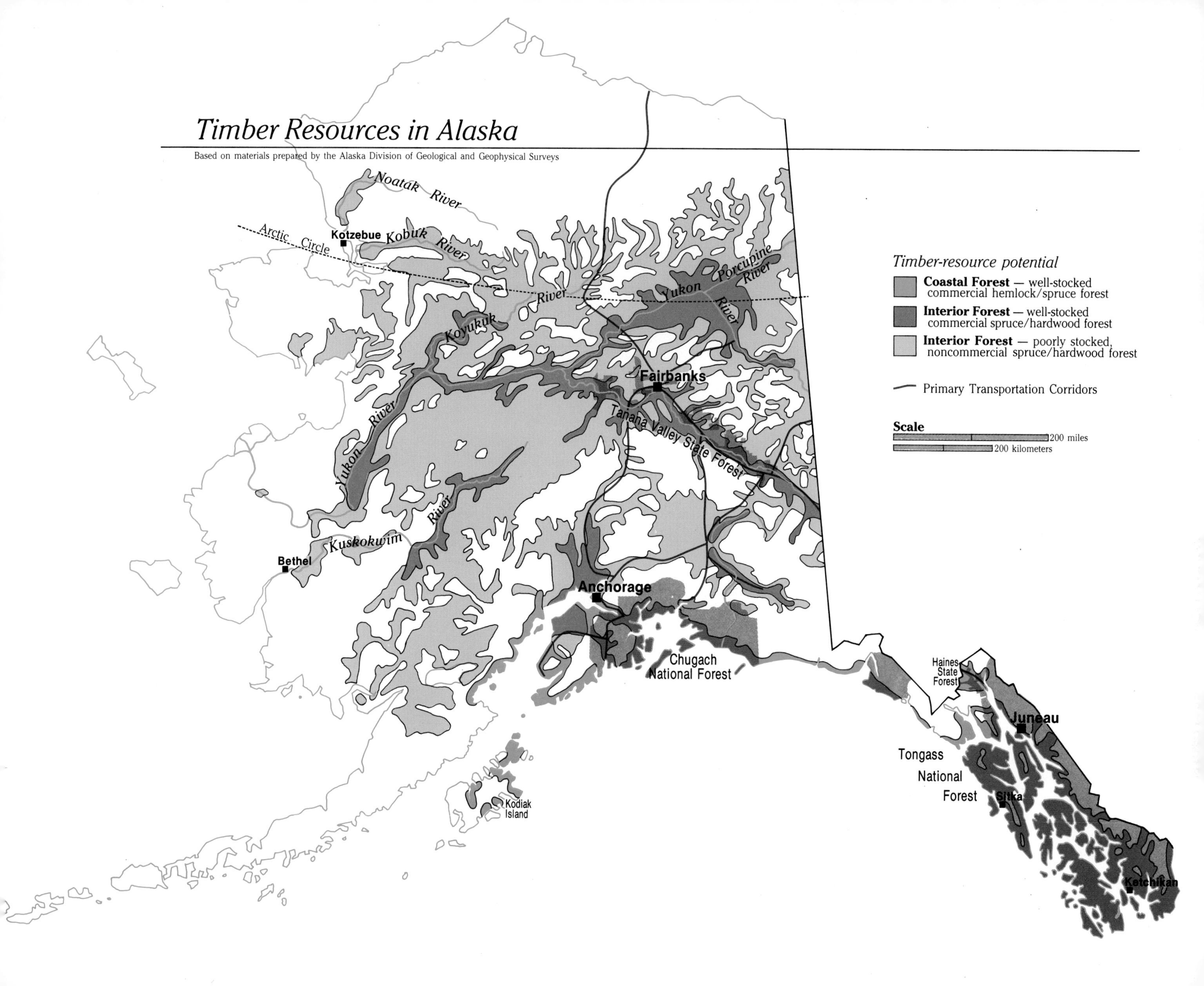

Timber Resources in Alaska
Based on materials prepared by the Alaska Division of Geological and Geophysical Surveys
Timber-resource potential
Coastal Forest — well-stocked commercial hemlock/spruce forest
Interior Forest — well-stocked commercial spruce/hardwood forest
Interior Forest — poorly stocked, noncommercial spruce/hardwood forest
Primary Transportation Corridors
Scale
200 miles
200 kilometers
Noatak River
Arctic Circle
Kotzebue
Kobuk River
Koyukuk River
Yukon River
Porcupine River
Fairbanks
Tanana Valley State Forest
Kuskokwim River
Bethel
Anchorage
Chugach National Forest
Kodiak Island
Haines State Forest
Juneau
Tongass National Forest
Sitka
Ketchikan

A small stand of balsam poplar, often confused with cottonwood, grows along the Noatak River. The species is common throughout the Interior, and is found as far north as the Arctic Slope.
(Lisa Holzapfel)

The farthest-north white spruce along the trans-Alaska pipeline grows near the Dalton Highway, four miles south of Chandalar Camp. (Charles Kay)

taiga. White spruce, principal commercial species in the Interior, abounds in the taiga. Paper birch, quaking aspen and balsam poplar accompany white spruce on warm, well-drained sites, while black spruce and tamarack (larch) occur on cold, wet sites. Willows provide food for wildlife. In a few locations, reindeer lichen and black spruce form communities important for caribou survival.

A continental climate, with temperatures which may vary 160° from summer to winter, gives rise to taiga. Daylight, almost nonstop in June, may be completely absent in December. Although annual precipitation rarely exceeds 20 inches, moisture from snowmelt is usually sufficient for spring growth. Rain during the growing season generally permits some forest growth despite the region's relatively cool conditions. Also important to the boreal forest is permafrost, perpetually frozen ground whose upper layer thaws every summer.

The Tundra

Tundra covers about half of Alaska, occurring west and north of tree line and also at altitudes too great for tree growth. The wide variety of tundra vegetation results from many differences in topography and soils.

The extremely short growing season — only 17 frost-free days at Barrow, for example — is somewhat offset by continuous daylight during the summer. Tundra plants often

Fast-growing alder is first to reclaim this landslide in southeastern Alaska. In 80 to 100 years, the alder will give way to spruce and hemlock. (Walt Matell)

Sitka spruce growing near Bering Glacier, along the Gulf of Alaska, act as soil traps, catching windblown soil and thereby improving the site for themselves and other plants. (Ken Winterberger)

resemble cushions that maximize volume and minimize surface exposure, thus conserving heat. Some cushion plants, called polsters, also have dark green leaves that absorb the sun's rays and raise the temperature inside the polster 25° to 40° warmer than the air.

Although little rain or snow falls on most of the tundra, permafrost blocks what water there is from penetrating more than a few feet. Stagnant pools warmed by continuous summer sun are splendid hatcheries for mosquitoes.

Strong winds are common on the tundra. To avoid being swept away or rasped by blowing grit, tundra plants must be well anchored and keep a low profile. Small, leathery or woolly leaves with rolled-under edges further reduce the impact of drying winds.

The Changing Forest

Disturbances to forest growth are normal, especially in the harsh subarctic environment. Some disturbances actually become requirements when they occur frequently enough to result in genetic and evolutionary changes.

Scientists are studying **ecological succession**, the way disturbed or bare ground is invaded by plant life. The nature of **succession** in the changing forest depends on the disturbance, the site and the plant communities involved.

Succession After Glaciation

Southeastern Alaska provides an excellent environment in which to study the gradual growth of new forests on raw land exposed by retreating glaciers.

In the Juneau area, willow, alder, cottonwood, Sitka spruce and hemlock grow on sites exposed for only two or three years. In Glacier Bay National Park and Preserve, less than 100 miles away, 40 to 50 years elapse before cottonwood, willow and spruce are firmly established, and 60 to 85 years must pass before alder takes hold. The time lag between the two areas is due to an abundant and diverse seed supply close to glaciers near Juneau, while at Glacier Bay the closest seed source is many miles away because the glaciers there are retreating more quickly.

At Glacier Bay, dryas (mountain avens) begins colonizing the fresh ground. Researchers believe that dryas helps pave the way for later plants by **fixing** atmospheric nitrogen and releasing it into the soil. Dryas is absent from early successional growth near Juneau, where its role is taken over by alder.

Fifty to sixty years after glacial retreat near Juneau, Sitka spruce overtops the alder and willow thickets and competes with black cottonwood. Decomposed litter from broadleaf trees contributes nitrogen and organic matter to the soil, thus aiding the spruce's rapid growth. As the spruce **canopy** closes, light-loving alder and willow weaken and die. The soil becomes more acidic and less fertile

A forest grows on once-bare land at the base of Black Rapids Glacier, in the Alaska Range south of Delta Junction. (Gary Michaelson)

as conifer needles make up a larger proportion of the debris on the forest floor. Western hemlock has adapted to grow in such an environment, and 200 years after glacial retreat hemlock achieves prominence in the **understory** of the spruce forest. At the same time, patches of sphagnum moss begin to take over on the ground. As the moss layer thickens, the forest floor becomes waterlogged and more acidic, with less available nitrogen. As larger trees die and fall, they create openings that let in some light, stimulating growth of some understory trees. The result is **old growth** forest.

But watch out for the mosses. On ground where drainage is poor, increased accumulations of organic matter can make the forest floor so soggy that even hemlock succumbs. Lodgepole pine can tolerate moist conditions and may survive on the edges of such muskegs. Sphagnum moss, sedges, rushes, lichens and leathery-leaved shrubs may become eventual heirs to the land uncovered by glaciers hundreds of years earlier.

Succession After Volcanic Eruption

Hundreds of miles to the west, in Katmai National Park and Preserve, another forest system recovers from disaster. Here, in 1912, a forest of balsam poplar, paper birch and white spruce was destroyed by volcanic eruption, earthquake, hurricane-force winds and burial under a thick layer of pumice and ash.

But succession started almost at once. Five years after the Valley of Ten Thousand Smokes was interred, a few faint traces of algae and moss existed around some of the cooler fumaroles. Two years later, the faint traces became dense patches. Apparently these simple plants were adapted to high temperatures, and could use ammonia gases from the steaming vents for their nitrogen requirements.

As the fumaroles cooled, plants capable of using atmospheric nitrogen started colonizing temperate surfaces. Half-inch-thick mats of liverworts carpeted large areas that had been barren 11 years earlier.

Today most of the valley's fumaroles are dead, and gone with them are the mosses, algae and liverworts that depended on their heat. The center of the valley is essentially unvegetated, for very little can survive the lunarlike conditions that exist there now. In some areas the sand has a thin, mossy film; in protected nooks, grasses and an occasional willow endure. Slopes and mountains surrounding the valley are more hospitable, and now support at least 35 different species of plants, including horsetail, fireweed, sorrel, poppy and willow. How many centuries will pass before the forest returns to these hills?

Succession After Fire

Fire is one of the most common occurrences on the taiga, but revegetation follows promptly. On warm sites, fast-growing

Skeletons of white spruce, killed by the 1976 eruption of Mount Augustine, are surrounded by new alder growth. (Don Cornelius)

Black spruce snags stand as reminders of wildfire that swept through this forest near Tok. (Walt Matell)

mosses, wildflowers, willow, birch and aspen reinvade the burned area within one to five years. Seven to thirty years later, solid willow thickets and birch and aspen begin to shade out the undergrowth. Eventually, more competitive trees dispatch the willow. In about 80 years, white spruce becomes conspicuous, and birch and aspen mature and then slowly decline. In some areas, these hardwoods can survive a century or longer in the white spruce forest that develops after a fire.

If fire doesn't set back the vegetation clock again, a white spruce stand with a thick moss mat eventually dominates. In colder areas, this may in turn be replaced by black spruce and bog, or a moss-lichen community, if permafrost develops.

Revegetation differs after fire on wet or permafrost-rich terrain where black spruce predominates. One to four years after the burn, bare mineral soil is covered with liverworts, mosses, fireweed and tree **seedlings**.

If the fire was light to moderate, low shrubs, horsetail and bluejoint grass sprout from **rhizomes**. The next stage is dominated by shrubs, but mosses, lichens and wildflowers increase as well. Twenty-six to fifty years after the fire, willow and alder thin out as black spruce overtops them. A black spruce forest develops, with its characteristic thick layer of ground moss. As moss builds up, the soil cools faster in autumn than it warms in spring, slowing release of nutrients and inhibiting root growth. Permafrost moves closer to the surface, further lowering soil temperatures and waterlogging the root zone.

During the summer the moss layer dries quickly, making the forest susceptible to wildfire. In the event it does not burn, black spruce and low shrubs continue to dominate the stand. Moss may cover most of the forest floor in a 100-year-old forest.

It is almost impossible to find a black spruce stand in Alaska more than 200 years old, so what eventually would develop is a puzzle. In wet lowlands, bog or a bog-forest cycle may replace mature black spruce. On better-drained uplands, black spruce may perpetuate itself by **layering** — growing vegetatively from buried branches.

Succession on Floodplains

Forest succession on newly formed river bars in the Interior differs from that after fire. No buried seeds or rhizomes are available to quickly regenerate the site. The whims of the

Succession on river bars takes place gradually, from a slow buildup of soil and silt to the establishment of seedlings. This photo shows shrubs and trees colonizing gravel bars in the Copper River near Gulkana. (Walt Matell)

Early successional hardwoods reclaim a tailings pile left behind by miners near the Chatanika River, northeast of Fairbanks. (Walt Matell)

river control early stages of forest development on a floodplain.

Floodplain succession begins when a sandbar is formed. In time, it may accumulate enough material to be above all but breakup and late-summer flooding. A salt crust can cover the surface until the bar is more than three feet above water level. Several years may pass before wind and water bring seeds of dwarf fireweed, horsetail, willow, balsam poplar and alder to new homes on this terrace. Once established, the plants act as barriers to salt crust formation, trap nutrient-rich silt carried downstream in spring floods, and stabilize the terrain. Alder is especially important for its ability to fix atmospheric nitrogen into the soil.

After five or ten years, one acre of river bar may contain 40,000 stems of alder and willow, producing more than two-and-one-half tons of wood and leaves per year. Life on the terrace is now firmly controlled by the plants, rather than the river.

Balsam poplar — black cottonwood in southcentral Alaska — also seed the new terrace. After 15 to 20 years, they overtop the thickets and shade them out. Falling leaves smother small undergrowth and add humus to the soil.

The terrace surface gradually rises above high water in 50 to 100 years, giving white spruce seeds flood-free periods in which to establish themselves.

The transition from hardwood to conifer

Jan Axell walks through a black spruce forest killed when the land along Turnagain Arm subsided during the 1964 earthquake. (Don Cornelius)

dominance takes about 75 to 125 years. Poplars, which must replace their entire set of leaves each year, are nutrient gluttons that depend on silt deposits to fertilize the soil. On islands and bars where heavy spring flooding is frequent, balsam poplar stands can last for several centuries.

Suffocating leaf-fall is reduced with the demise of the poplars, so mosses reappear in great numbers, in turn concealing the bare ground necessary for spruce regeneration. Soil temperatures decrease, leading to development of permafrost.

Growth of white spruce is usually greatly reduced after about 100 years. Black spruce, which can tolerate permafrost, may eventually replace white. In the absence of fire, the white spruce/black spruce/moss forest would eventually degenerate into a boggy woodland. If the insulating moss layer is disturbed, thawing of the permafrost causes the formation of small ponds, and gradual development of a treeless bog.

Tectonic Forest

We have seen how the forests of Alaska are shaped by glaciers, volcanoes, fires and rivers. Forests near the Copper River Delta have been shaped by another force: earthquake. Before A.D. 765, Sitka spruce covered the delta. Then, tectonic activity caused the land to sink and the ocean to drown the forest. The delta became a maze of brackish marshes, sloughs and ponds. Salt concentra-

A forester examines young Sitka spruce growing along a recently uplifted stretch of beach near Cape Yakataga on the Gulf Coast. (Ken Winterberger)

These 80-year-old spruce are stunted along the Noatak River, northern limit for the species.
(Lisa Holzapfel)

tion, soil drainage and tidal flooding determined what plant grew where.

On Good Friday 1964, the delta was wracked again. This time the land was uplifted about six feet, draining ponds and elevating sloughs above tidewater.

Thus, the delta may now be returning to the conifer forests of 1,200 years ago. As a result of the upheaval, vegetation requiring wet soil is decreasing, and plants preferring better-drained habitats are expanding. Salt is leaching from the soil, allowing a greater variety of plants to gain a foothold. Sedges are invading the margins of sunken ponds, dry channels and uplifted flats. Former sedge territory is now being invaded by grasses, herbaceous plants and shrubs. Alder and willow are expanding seaward on banks of sloughs and well-drained surfaces. Perhaps these developments represent the early successional shrub stage that frequently heralds the arrival and eventual dominance of tree species.

The Environmental Sieve

Climate and fire are examples of external forces shaping growth and development of forests. Internal forces such as birth, death and competition between individuals and species are also necessary in this ecological game, in which survivors of the environmental sieve pass on successful characteristics. But the game has no end, because the rules and players are forever changing.

Alaska's Trees

[**Editor's note:** *The following pages describe Alaska's tree species. For a more comprehensive treatment, we recommend the book from which much of this chapter was derived:* Alaska Trees and Shrubs, *by Leslie A. Viereck and Elbert L. Little, Jr., U.S. Department of Agriculture Handbook Number 410 (1972). Species sketches and distribution maps are courtesy of the U.S. Forest Service.*]

A **tree** is defined as a woody plant at least 12 feet tall, with a trunk at least 3 inches in diameter, and a **crown** of foliage. Twelve Alaskan tree species can exceed 70 feet, five are between 30 and 70 feet and sixteen are less than 30 feet in height. Alaska has only 33 native tree species, fewer than any of the other states.

Many of Alaska's trees range into northern Canada, and even to the Lower 48. One species, Sitka alder, is also native to eastern Siberia.

Western Hemlock (figure 1)

(Tsuga heterophylla)
Also called: Pacific hemlock, west coast hemlock.

About two-thirds of southeastern Alaska's forest is western hemlock, one of the state's principal commercial species. Common size at maturity is 100 to 150 feet tall, with trunk diameters of 2 to 5 feet. Some trees may be more than 1,000 years old.

Hemlock tolerates shade well, and is able to grow slowly under the dense canopies found in coastal forests. The tree ranges from the Kenai Peninsula to southeastern Alaska, and is also found in British Columbia, the Pacific Northwest, northern California and northwestern Montana.

Hemlock is made into dissolving pulp, but **lumber** quality logs are also harvested. In some stands, trees suffer from **fluting**,

Early morning mist settles among the trees of a spruce forest in the Kenai National Wildlife Refuge.
(Rollo Pool)

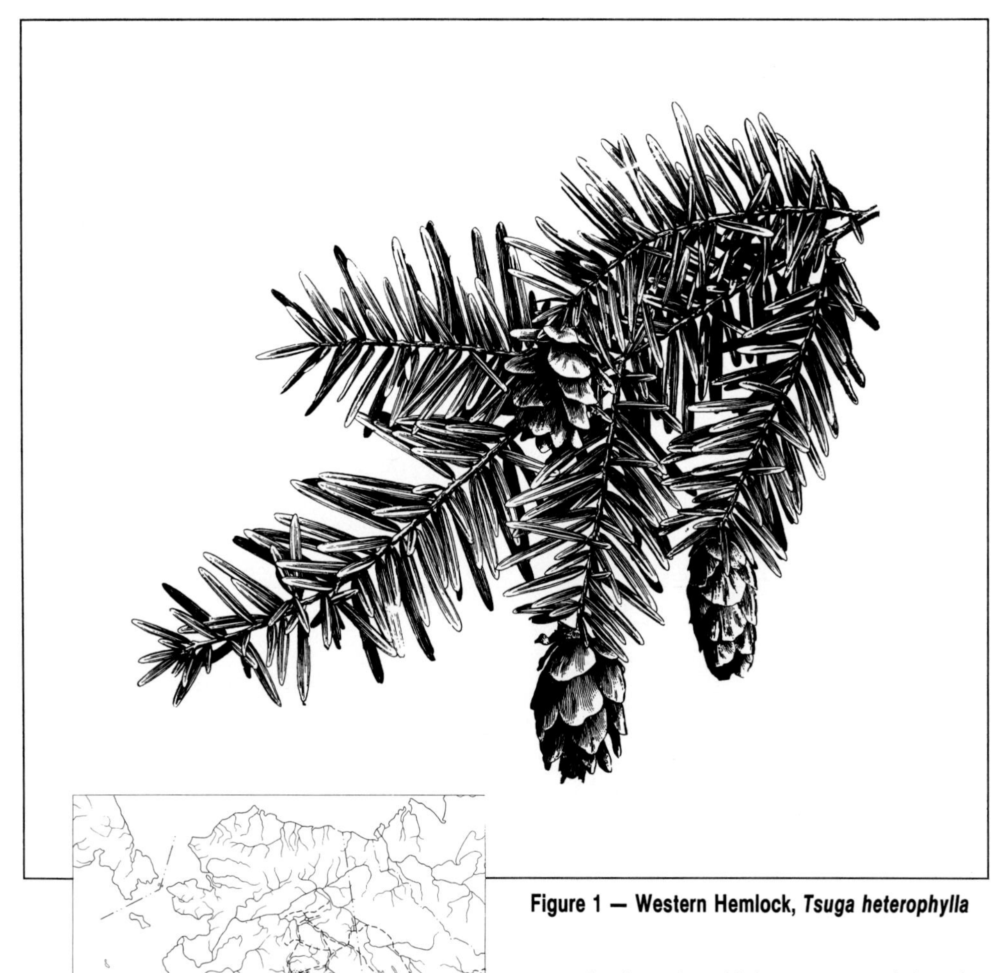

Figure 1 — Western Hemlock, *Tsuga heterophylla*

growth of trunks with buttresses, a defect that makes them unsuitable for lumber.

What to look for: Needles, rounded and flexible, ¼ to ⅞ inch long, shiny dark green on top, with two whitish lines on the bottom surface. Twigs flexible, slender, with fine hairs and rough peglike bases where old needles used to be. Bark reddish to gray-brown, with scaly plates; cut with knife to reveal red inner bark. Cones, ⅝ to 1 inch long, hang down from ends of branches. The very top twig of a hemlock, called the leader, is often droopy.

Sitka Spruce (figure 2)

(Picea sitchensis)

Also called: tideland spruce, yellow spruce, western spruce or coast spruce.

The official state tree, Sitka spruce, can be 750 years old, reach heights up to 225 feet, and have trunk diameters exceeding 10 feet. Commonly between 100 and 160 feet high, with 3- to 5-foot diameter trunk.

Ranges the coastal forests from the Kenai Peninsula to northern California. About one quarter of the forest in southeastern Alaska is Sitka spruce; on Afognak and Kodiak islands, it is the only conifer. The species usually grows from sea level to 1,500 feet, but dwarfed spruce have been recorded on 3,500-foot nunataks above the Juneau Icefield.

Sitka spruce is a prolific seed producer. In windswept areas where the tree grows in a bushy form low to the ground, the spruce also propagates by layering: branches that touch the ground get covered up, and eventually develop roots and a new stem.

Growth rings in old-growth spruce are usually closely spaced, producing a strong, fine-grained wood. Such trees are quite valuable, and have been selectively logged

Figure 2 — Sitka Spruce, *Picea sitchensis*

from the coastal forests for many years. Spruce's high ratio of strength-to-weight made the wood desirable for aircraft construction during the first and second World Wars.

Today, fine-grained spruce logs are made into piano and guitar sounding boards, gliders, boats and scaffolding, and are used in construction. Lower grades can be pulped.

What to look for: Needles stiff, sharp, ⅝ to 1 inch long, stand out on all sides of the twigs. Bark thin, gray and smooth, becoming dark purplish-brown with scaly plates on older growth. Cylindrical hanging cones are 2½ to 3½ inches long; ½-inch-long seeds have wings.

Western Redcedar (figure 3)

(Thuja plicata)

Also called: canoe cedar, shinglewood, Pacific redcedar, arborvitae.

The southern half of the Alaska Panhandle marks the northern limit of Western redcedar, which ranges south along the coast to California and into Montana. In Alaska, the tree can grow to more than 150 feet tall, with tapered trunks on buttressed bases 9 feet in diameter. Commonly 70 to 100 feet, with 2- to 4-foot trunks. Trees 800 years old have been reported.

Aromatic and quite resistant to decay, redcedar wood has traditionally been used by coastal Natives to make totem poles and canoes. Today the species is used for shingles and shakes, utility poles and light construction. The wood can be pulped by the kraft process, a method not used in Alaska today.

In the past, stands with high cedar content were not harvested, or the cedar was cut but left in the woods. Today, however, good quality logs command high prices on the Japanese export market.

What to look for: Leaves scalelike, 1/16 to ⅛

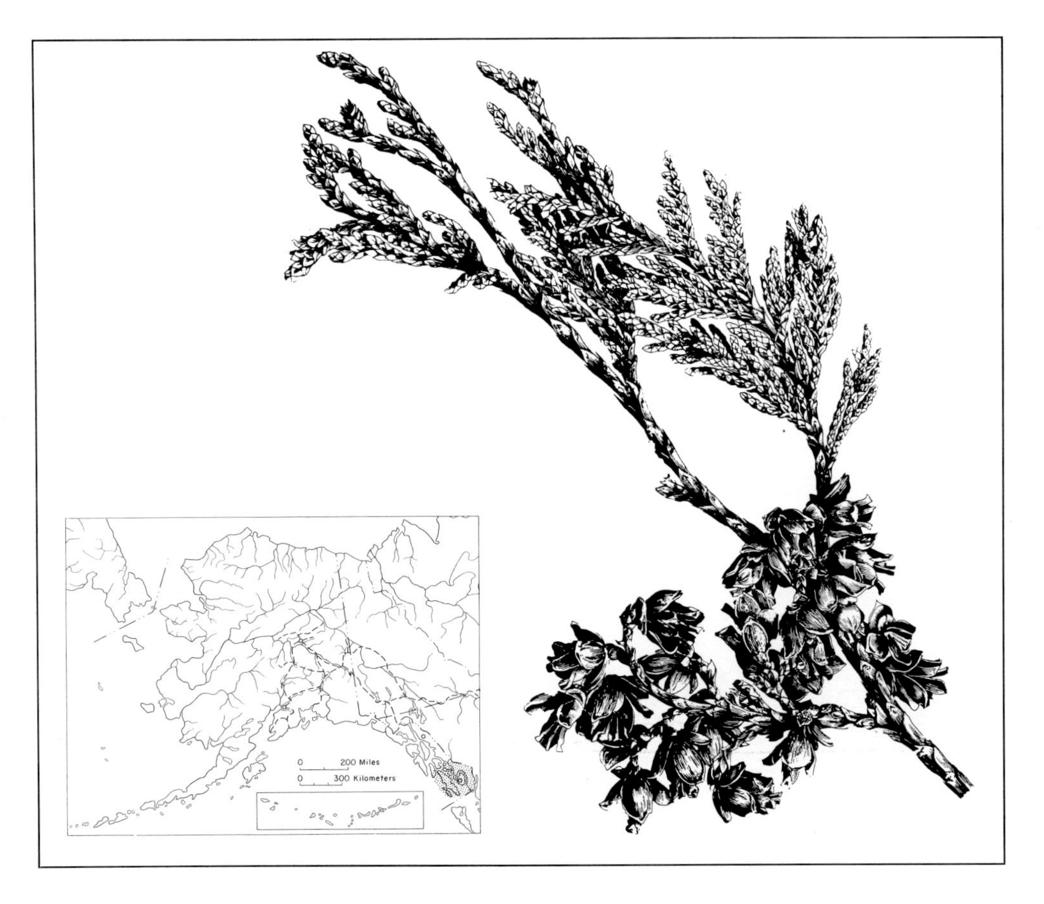

Figure 3 — Western Redcedar, *Thuja plicata*

inch long, shiny yellow-green tops and dull green bottoms. Leafy twigs, flattened as though pressed by an iron, in slightly drooping fanlike sprays. Older twigs gray and smooth. Bark thin, gray or brown, fibrous and shreddy, becoming thick and furrowed into long ridges in older growth. Wood has distinctive cedar odor. Heartwood is reddish-brown. Light brown cones are elliptical, ½ inch long, turned up on short stalks, clustered near ends of twigs.

Alaska-Cedar (figure 4)

(Chamaecyparis nootkatensis)

Also called: yellow cedar, Alaska yellow-cedar, Alaska cypress, Nootka false-cypress, yellow cypress.

This slow-growing cedar typically reaches 40 to 80 feet in height, with trunk diameters 2 feet or less. Notable specimens have reached 120 feet tall with 8-foot diameters. The species is scattered from Prince William Sound south along the coast to northern California, with concentrations on Baranof and Chichagof islands south of Glacier Bay.

Traditional native canoe paddles were made from Alaska-cedar. The durable, aromatic wood is used for boat building, windows, exterior doors, and furniture. The wood is prized in Japan; prices can exceed one dollar per board foot in the uncut round log.

What to look for: Wood has a distinctive odor. Leaves scalelike, 1/16 to 1/8 inch long, yellow-green, with slightly spreading tips, on leafy twigs that are flattened or squarish in cross-section. Flat, slightly drooping spreading sprays. Bark gray or brown, with long, narrow shreds and fissures. Heartwood is bright yellow. Round cones, less than ½ inch in diameter, scattered (not clustered) on sprays.

Mountain Hemlock (figure 5)

(Tsuga mertensiana)

Also called: black hemlock, alpine hemlock.

Closely related to western hemlock, this species occupies a similar range, although it is less common in Alaska. Mountain hemlock can grow at higher elevations; above timberline, it is stunted to dwarf ***krummholz***, low-lying shrubs. Typical size of the erect form in Alaska is 50 to 100 feet tall, with a 2- to 3-foot-diameter trunk.

When mixed with western hemlock in commercial stands, mountain hemlock is also

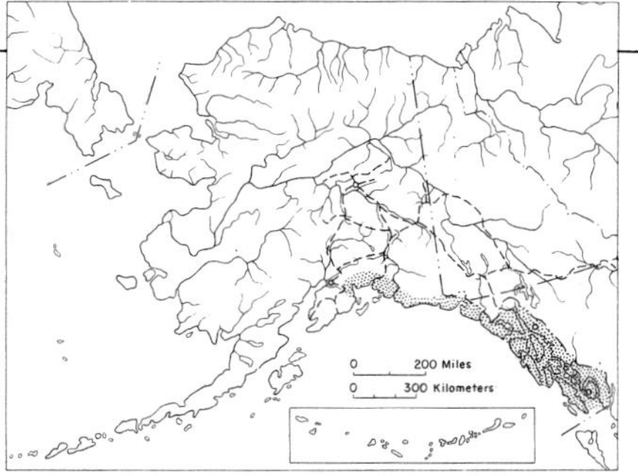

Figure 4 — Alaska-cedar, *Chamaecyparis nootkatensis*

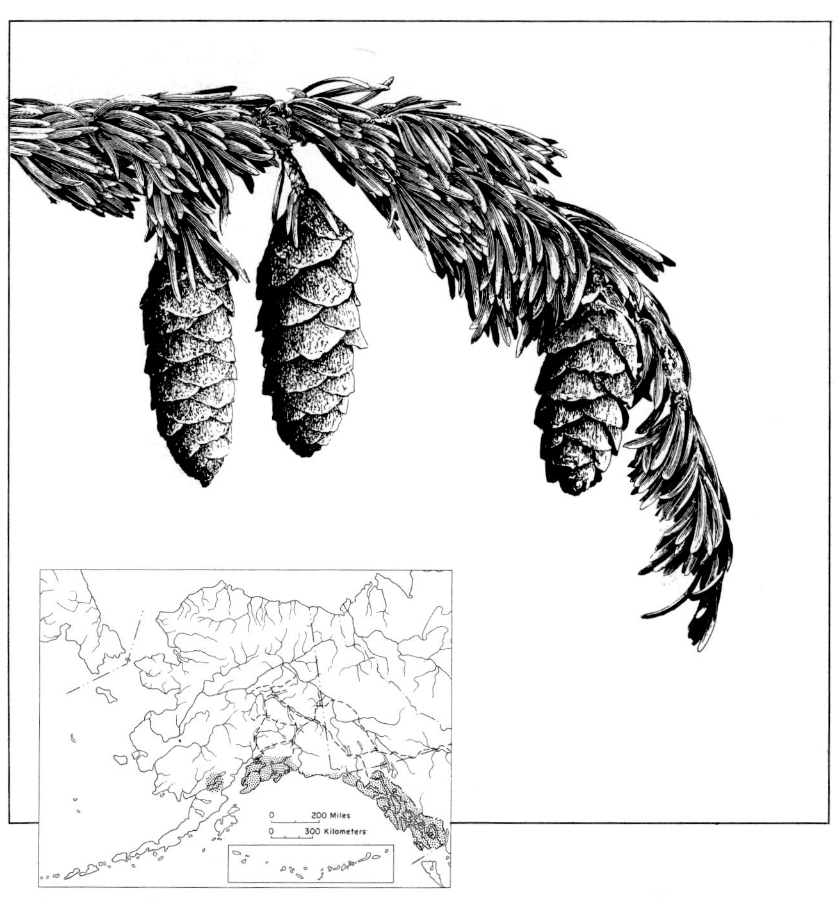

Figure 5 — Mountain Hemlock, *Tsuga mertensiana*

logged with no attempt to separate the species. Mountain hemlock wood is slightly more dense than western hemlock, makes good pulp, and is used in similar ways. As logging moves onto poorer sites and higher elevations, mountain hemlock will become more important commercially.

What to look for: Needles short-stalked and blunt, ¼ to 1 inch long, blue-green with whitish lines, crowded on all sides of short side twigs. Twigs short, slender, light reddish-brown, with peglike bases where needles formerly grew. Bark thick, gray to dark brown, deeply furrowed into scaly plates. Pale red-brown heartwood; similar or lighter-colored sapwood. Stalkless cylindrical cones, 1 to 2½ inches long, hanging down, purple when fresh but turning brown. Half-inch seeds have long wings.

White Spruce (figure 6)

(Picea glauca)

Also called: western white spruce, Canadian spruce, Alberta spruce, Porsild spruce.

White spruce is the most abundant and most important commercial tree species of interior Alaska. Commonly 40 to 70 feet high, with 6- to 18-inch trunk diameters; on good sites white spruce can grow to 115 feet with 30-inch-diameter trunks.

White spruce is a favorite for house logs, because of its strength, light weight and

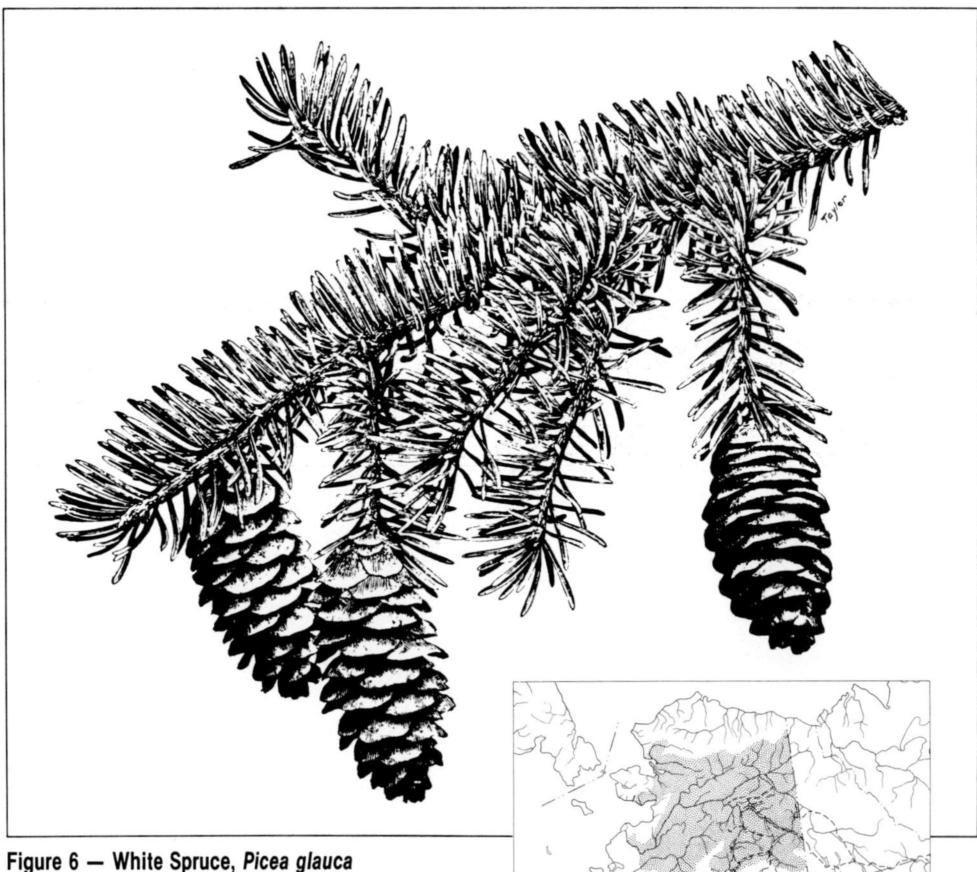

Figure 6 — White Spruce, *Picea glauca*

favorable sizes. The species ranges from the Chugach Mountains to the Brooks Range, and extends into Canada to the east.

White spruce in Alaska are more slender and spirelike than varieties found in eastern Canada and the Great Lakes States. At timberline, the tree can become a prostrate shrub.

What to look for: Needles sharp, stiff, ½ to ¾ inch long, blue-green with whitish lines on all sides, growing on all sides of twigs but concentrated on top near the ends. Twigs hairless, orange-brown, with rough peglike bases where needles formerly grew. Bark thin, gray, smooth or in scaly plates; inner bark whitish. Wood almost white. Cones stalkless, cylindrical, 1¼ to 2¼ inches long, shiny light brown with thin, flexible scales. Mature cones fall off tree.

Black Spruce (figure 7)

(Picea mariana)

Also called: bog spruce, swamp spruce.

Most black spruce grow to 15 to 30 feet, with trunk diameters between 3 and 6 inches, although one 72-foot specimen has been reported.

Black spruce ranges between the Chugach Mountains and the Brooks Range. The species favors cold, wet areas like north-facing slopes, muskegs, valley terraces and lake margins, and can tolerate permafrost better than other tree species.

What to look for: Needles pointed, ¼ to ⅝ inch long, ashy blue-green with whitish lines on all sides, spreading on all sides of the twig. Twigs slender, covered with short, reddish hairs, becoming brown and rough from peglike bases where needles grew. Bark thin, composed of gray or dark gray scales, brown underneath; inner bark yellowish. Light yellow wood has very fine growth rings, attesting to the species' slow growth. Cones round or egg-shaped, ⅝ to 1¼ inch long, gray or black, curving downward on short stalks. Mature cones may remain closed, in clusters near treetops, for several years.

Figure 7 — Black Spruce, *Picea mariana*

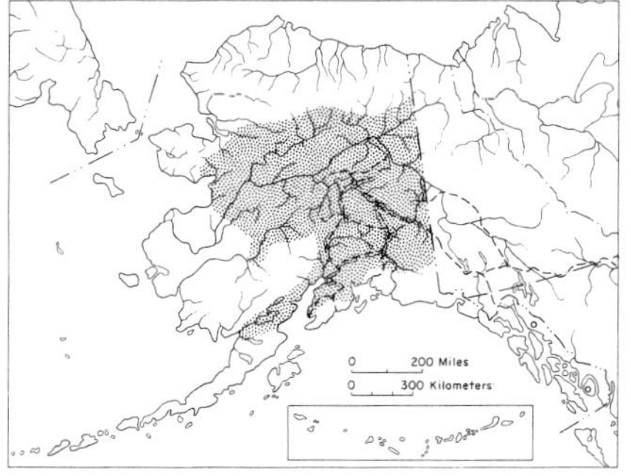

Paper Birch (figures 8a, 8b and 8c)

(Betula papyrifera)

Also called: white birch, canoe birch.

Paper birch, one of North America's most widely distributed trees, ranges from Pennsylvania and Iowa up through Canada and across Alaska's Interior to the Seward Peninsula. In Alaska, it grows to medium size, commonly 20 to 60 feet tall, with a 4- to 12-inch trunk diameter.

Alaska has three varieties of the species,

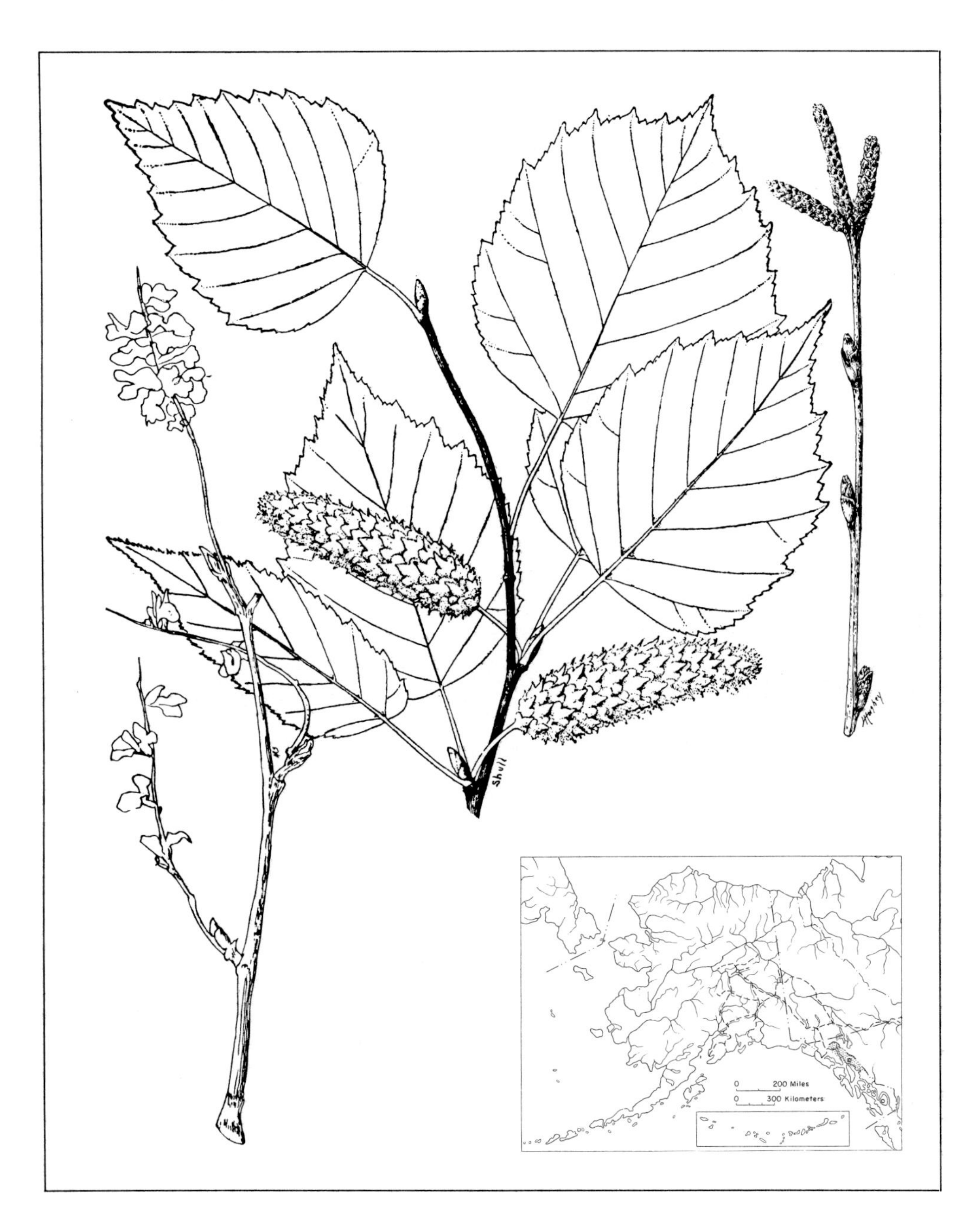

Figure 8a — Western Paper Birch, *Betula papyrifera* var. *commutata*

which hybridize where their ranges overlap: Alaska paper birch (variety *humilis*), throughout the Interior; western paper birch (variety *commutata*), southeastern Alaska from the Taku River to Lynn Canal; and Kenai birch (variety *kenaica*), found on the Alaska Peninsula, Kodiak Island, Kenai Peninsula, southcentral Alaska and parts of the Interior. Paper birch will also hybridize with shrublike dwarf arctic (alpine) birch (*Betula nana*) and resin birch (*Betula glandulosa*).

Natives have traditionally used the tree's smooth bark for canoes and basketry. Only a small amount is harvested for lumber; the wood can be used for furniture, toys, handles, toothpicks, veneer and for carving. Birch is also cut for fuelwood, and some people tap its sap to make birch syrup.

What to look for: Leaves rounded or wedge-shaped at base and pointed at tip, on slender stalks ½ to 1 inch long. Leaf blades 1½ to 3½ inches long, with coarse teeth along edges, dull dark green and smooth on top, yellow-green with perhaps some hair on bottom. The most obvious and striking feature of paper birch is its bark, which peels off easily into thin strips. Mostly white, bark can also be pinkish, coppery-brown or purplish-brown; inner bark is orange. Male and female flowers on same twig. Fruits cylindrical, look like soft cones, 1 to 2 inches long, ⅜-inch in diameter, containing numerous tiny seeds (nutlets) with wings.

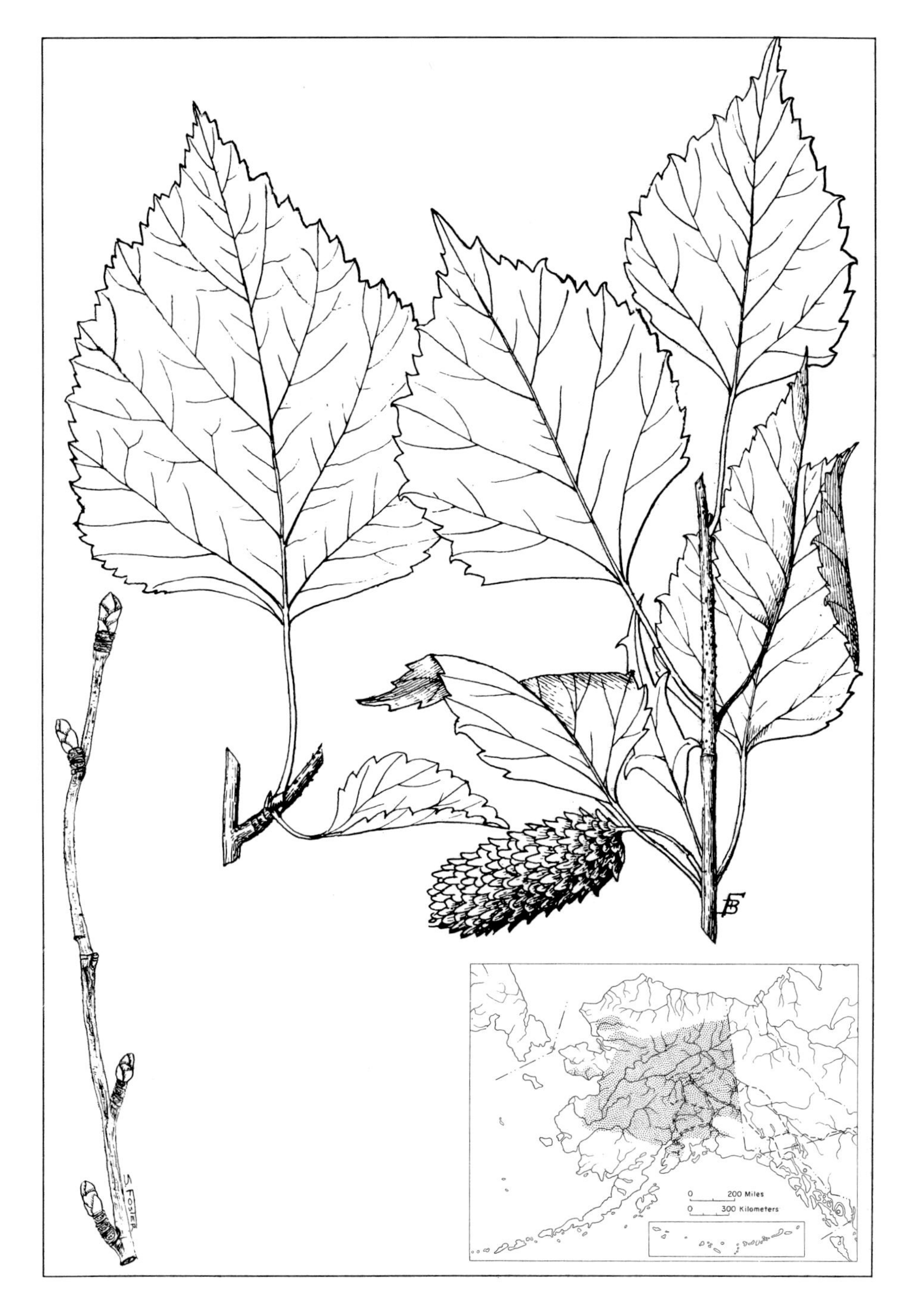

◄**Figure 8b — Alaska Paper Birch, *Betula papyrifera* var. *humilis***

▼**Figure 8c — Kenai Birch, *Betula papyrifera* var. *kenaica***

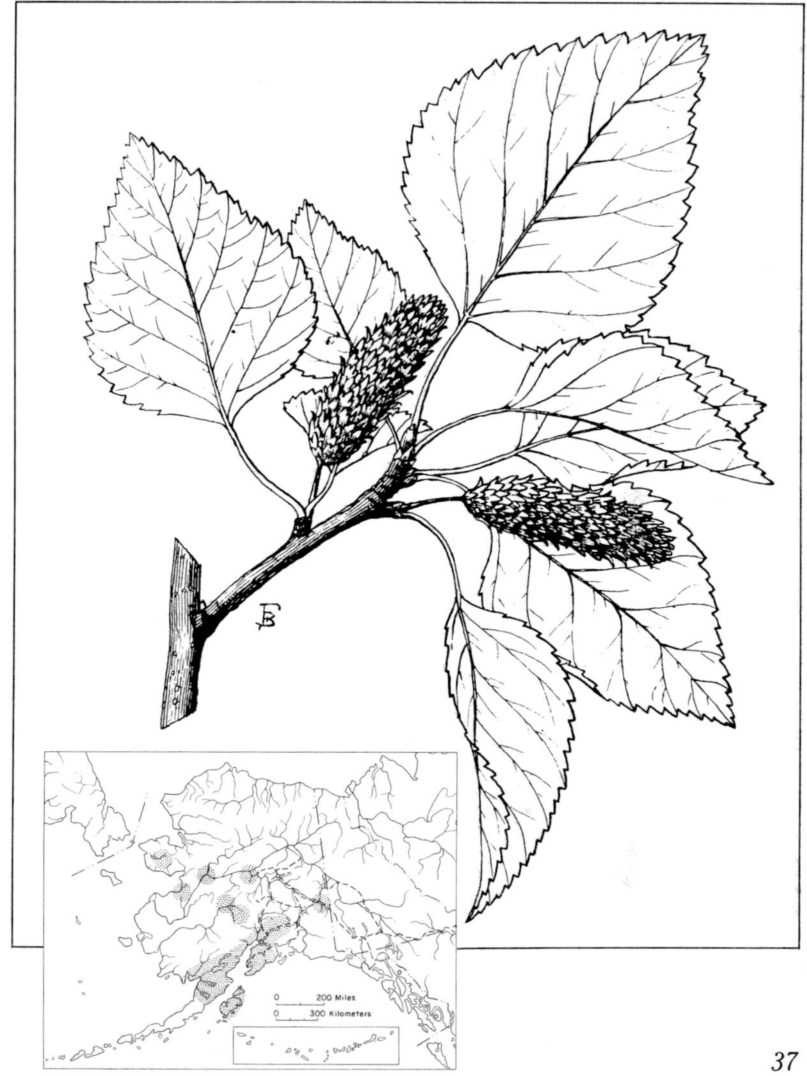

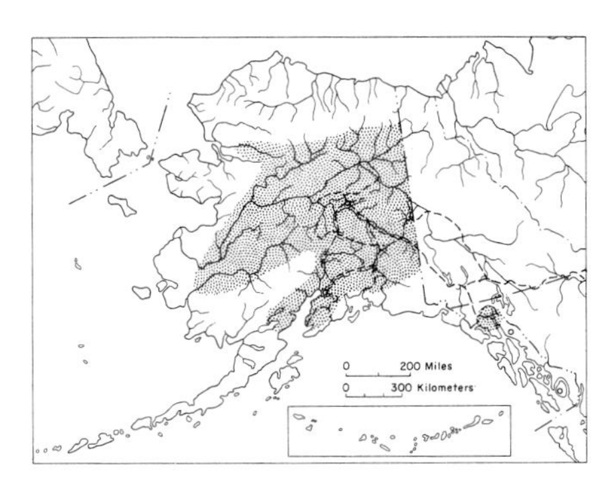

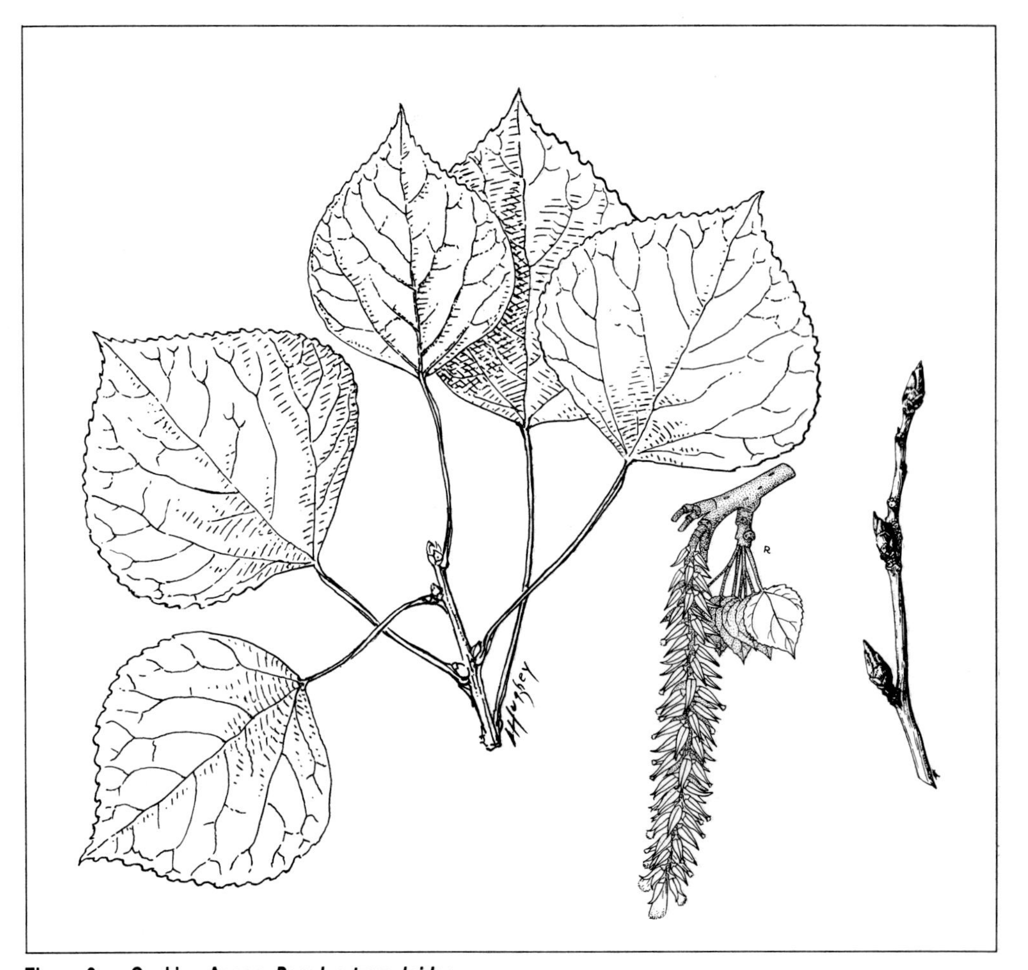

Figure 9 — Quaking Aspen, *Populus tremuloides*

Quaking Aspen (figure 9)

(Populus tremuloides)

Also called: American aspen, trembling aspen, popple.

This fast-growing tree commonly occupies south-facing slopes, well-drained benches and creek bottoms throughout the Interior and near the head of Lynn Canal in southeastern Alaska. Trees are usually 20 to 40 feet tall, with 3- to 12-inch-diameter straight trunks.

Male and female **catkins** grow on different trees. After fire, aspen can propagate rapidly from root suckers, often resulting in pure stands that continue to grow for up to 100 years.

Although not used commercially in Alaska to any extent, aspen has been processed into pulp, or cut into lumber, matches and packing excelsior in Canada and the Lower 48. It can also be chipped and compressed into wafer board.

What to look for: Leaf blades, 1 to 2 inches long, nearly round with short points, are attached to twig by stalks 1½ to 2½ inches long, flattened so they flex easily in one direction (this enables even a slight wind to flutter the leaves; hence the tree's name). Twigs slender, reddish and slightly hairy when young, becoming gray with raised leaf scars that show three dots on older growth. Bark whitish or green-gray, smooth with curved scars and black knots. Drooping catkins, 1½ to 2½ inches long, contain many individual flowers and blossom in May before leaves emerge. Tiny cottony seeds are released to the wind in summer.

Balsam Poplar (figure 10)

(Populus balsamifera)

Also called: tacamahac, cottonwood, tacamahac poplar, balm of Gilead (erroneously).

The Interior's largest hardwood, balsam poplar occasionally reaches 100 feet with 2-foot-diameter trunks. It commonly grows to 30 to 50 feet, with 4- to 12-inch trunks.

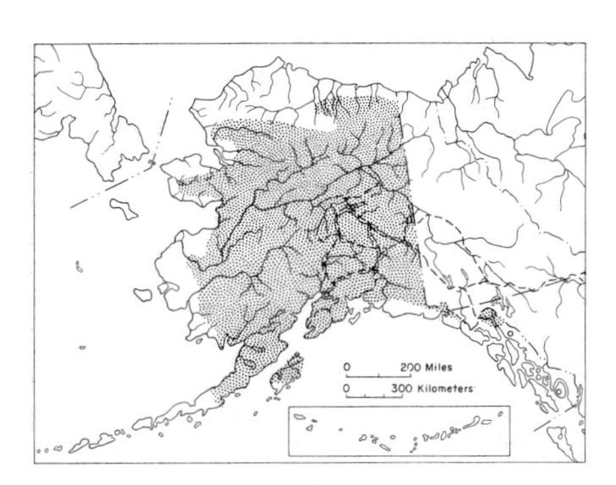

Balsam poplar grows in river valleys throughout the Interior, in northern southeastern Alaska, in the Brooks Range and even on the Arctic slope. It can be found at elevations up to 3,500 feet. Where its range overlaps black cottonwood, the two species hybridize. Poplar will on rare occasions also hybridize with quaking aspen.

In Alaska, balsam poplar is used primarily for fuelwood and specialty lumber. Elsewhere, it is used for crates, boxes and pulpwood.

What to look for: Leaf blades, shaped somewhat like arrowheads, 2½ to 4½ inches long, attached to twigs by slender 1- to 2-inch-long stalks. Leaves shiny dark green on top, pale green and rusty brown on the bottom. Twigs red-brown and hairy, with orange dots when young; older twigs become gray with raised leaf scars showing three dots. Bark light gray, initially smooth, becoming rough, thick and furrowed. Male and female catkins, 2 to 3½ inches long, hang on separate trees. Capsules in mature female catkins release many tiny cottony seeds in mid-summer. Sometimes mistaken for black cottonwood.

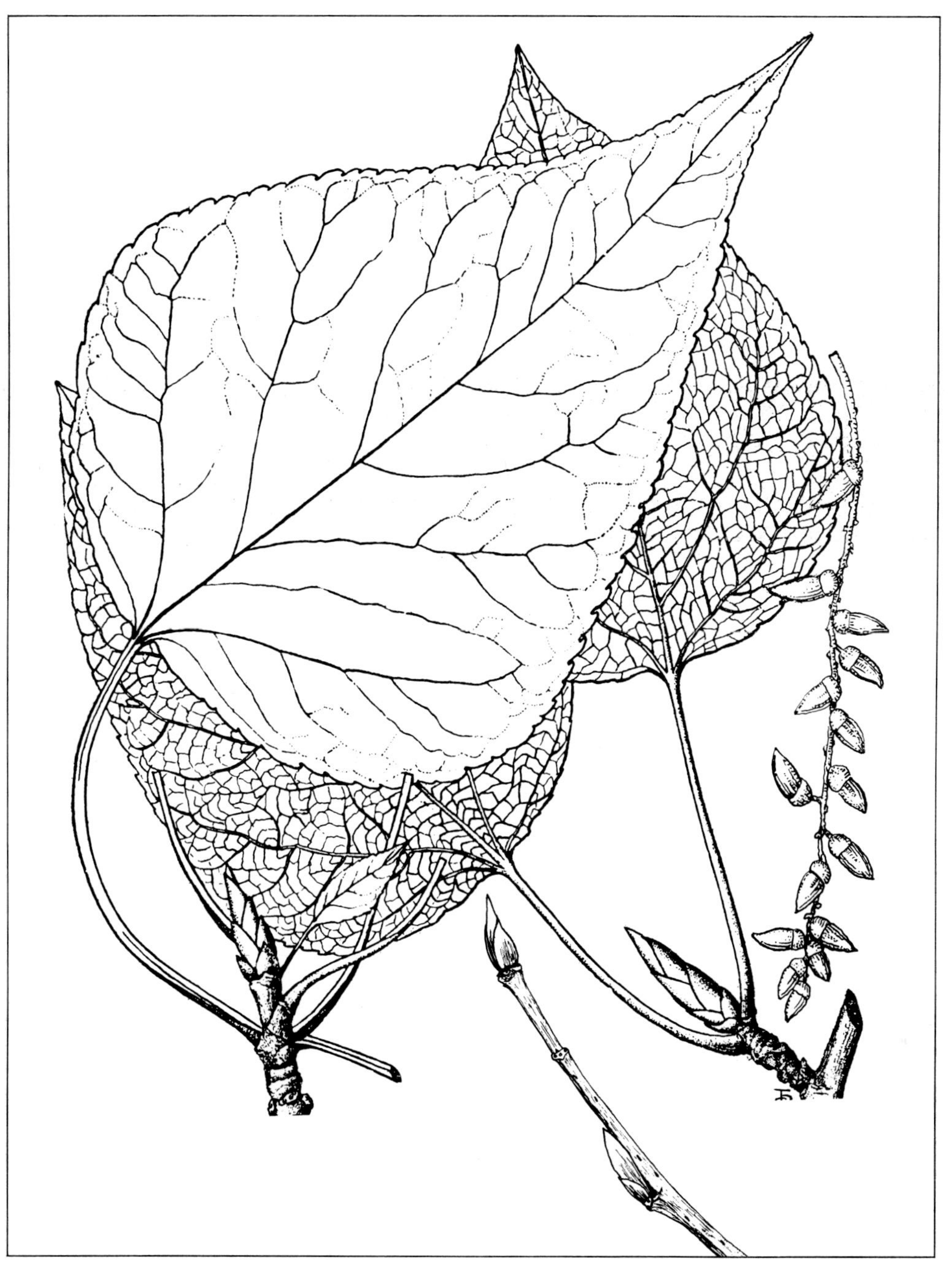

Figure 10 — Balsam Poplar, *Populus balsamifera*

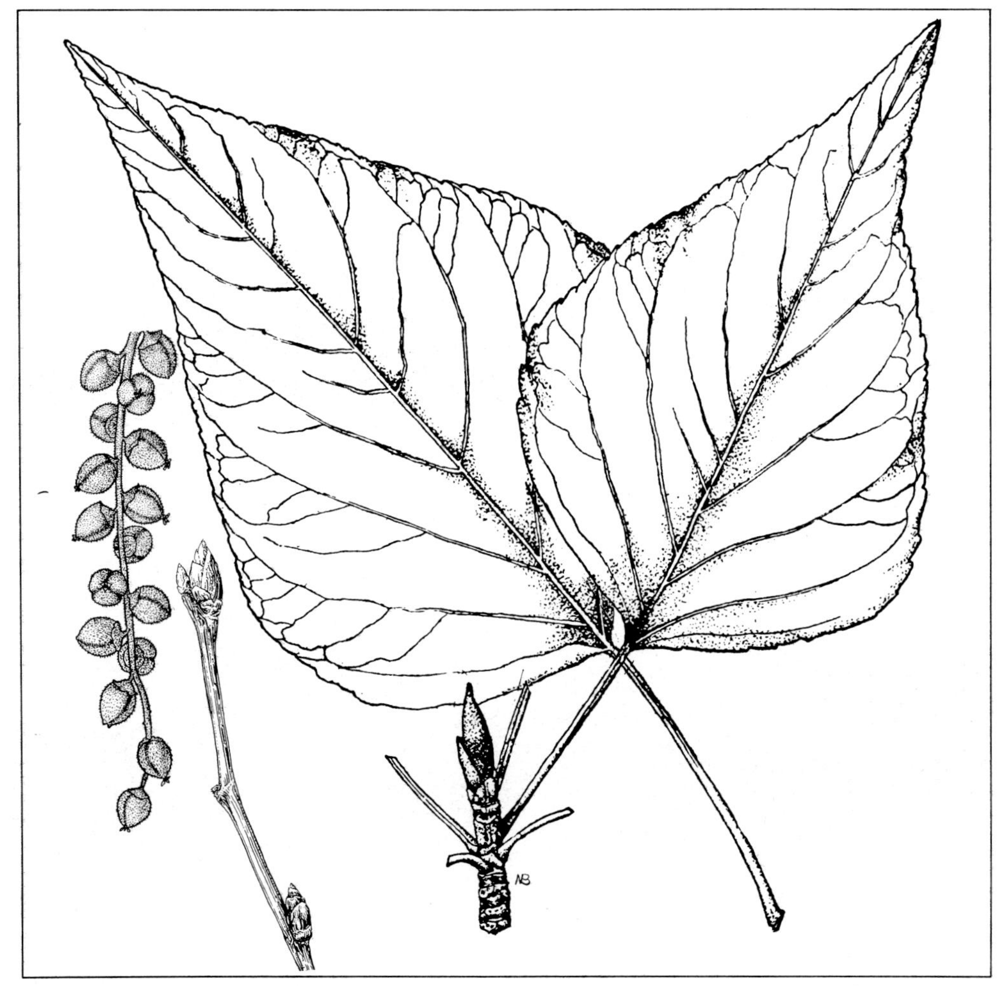

Figure 11 — Black Cottonwood, *Populus trichocarpa*

Black Cottonwood (figure 11)

(Populus trichocarpa)

Also called: balsam cottonwood, Pacific poplar, northern black cottonwood.

Largest broadleaf tree species in Alaska, black cottonwood rapidly attains heights of 80 to 100 feet, with 30-inch-diameter, straight trunks.

Black cottonwood grows in lowlands, large river valleys and glacial outwash plains of southern and southeastern Alaska. Its wood has been used to a limited extent for house logs, lumber and pulp. The species is often grown as an ornamental or shade tree.

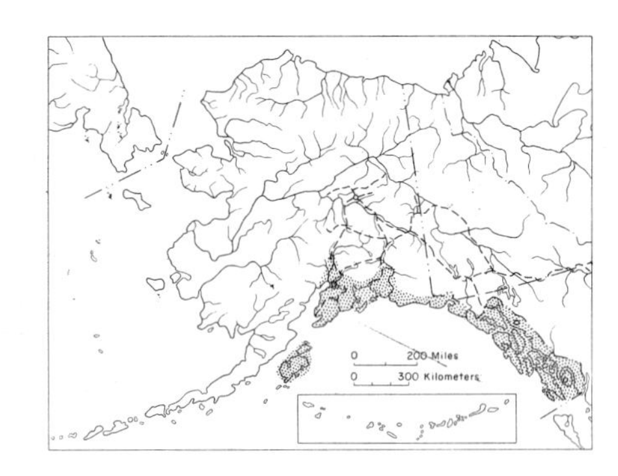

Black cottonwood can be mistaken for its close relative, balsam poplar, whose range it overlaps in the Cook Inlet and upper Lynn Canal areas. Leaves of black cottonwood are generally proportionately broader and whiter on the bottom than poplar. Cottonwood seed capsules are nearly round, hairy, and split into three parts; poplar seed capsules are elongated, pointed, hairless, and split into two parts.

What to look for: Leaf blades 2½ to 5 inches long, shaped like arrowheads, with finely toothed margins, shiny dark green on top, whitish with rusty specks on bottom, on round stalks measuring 1½ to 2 inches. Twigs red-brown and hairy when young, becoming dark gray with raised leaf scars showing three dots. Bark gray to dark-gray, smooth, becoming thick, round and furrowed with age. Catkins 1½ to 3 inches long, drooping with many ⅛-inch flowers; male and female flowers on separate trees. Cottony seeds released in mid-summer.

Figure 12a — Shore Pine, *Pinus contorta* var. *contorta*

Figure 12b — Lodgepole Pine, *Pinus contorta* var. *latifolia*

Lodgepole Pine (figures 12a and 12b)
(Pinus contorta)
Also called: shore pine, scrub pine, tamarack pine.

There are two varieties of Alaska's only native pine species: shore pine (*Pinus contorta* var. *contorta*), found throughout the Panhandle; and the regular inland variety (*Pinus contorta* var. *latifolia*), found near Skagway and Haines.

The inland variety spills across the Coast Range from the northern Canadian Rockies. It grows to 75 feet, with trunk diameters to 12 inches. Its hard, heavy cones remain closed for years on the tree. Inland pine is a commercial species in the Rockies.

The shore pine variety commonly grows in

and around moist muskegs throughout southeastern Alaska, where it usually reaches 20 to 40 feet in height. Occasionally it will grow to 75 feet, with trunk diameters up to 32 inches. On the poorest sites, the pine becomes a shrub.

What to look for: Needles, 1 to 2¼ inches long, two to a bundle, yellow-green to dark green with whitish lines. Stout twigs, orange when young, becoming gray and rough with age. Bark, gray to dark brown, scaly. Resinous wood with spiral grain in scrubby trees. Egg-shaped cones, 1¼ to 2 inches long, can remain on tree for many years.

Tamarack (figure 13)

(Larix laricina)

Also called: Alaska larch, eastern larch, hackmatack.

Tamarack is the only Alaskan conifer to shed its needles yearly. Commonly 30 to 60 feet tall, with 4- to 10-inch-diameter trunks; tamarack occasionally grows to 75 feet with 13-inch trunks.

The species abounds along the Tanana, Yukon, Kuskokwim and Koyukuk rivers. Its durable wood is used for fence posts, railroad ties and poles.

What to look for: Crowded clusters of 12 to 20 narrow and flexible needles, ⅜ to 1 inch long, blue-green turning to yellow before falling off in autumn. Twigs stout, hairless, with short spur twigs that hold the needles.

Figure 13 — Tamarack, *Larix laricina*

Bark dark gray, thin and smooth, becoming scaly in older growth, exposing brown wood beneath. Cones round, ⅜ to ⅝ inch long, turned up on horizontal twigs, opening in early autumn but remaining on trees through the winter.

Subalpine Fir (figure 14)

(Abies lasiocarpa)

Also called: alpine fir, white fir.

This conifer, rare in Alaska except in a few locations, grows in some valleys and mountains of southeastern Alaska. Commonly 20 to 60 feet tall, with 4- to 12-inch-diameter trunks; occasionally to 95 feet with 27-inch trunks. Near tree line, it takes a prostrate shrublike form. The species ranges from Yukon Territory, south through southeastern Alaska and British Columbia to the mountains of New Mexico and Arizona.

No commercial use for this tree is recorded in Alaska.

What to look for: Needles ¾ to 1¼ inches long, dark blue-gray, crowded on gray, rusty, hairy twigs. Bark thin, smooth, gray. Cylindrical cones 2½ to 4 inches long, sitting upright on top branches of tree. In autumn, cones disintegrate, releasing seeds; whole cones are not found on the forest floor.

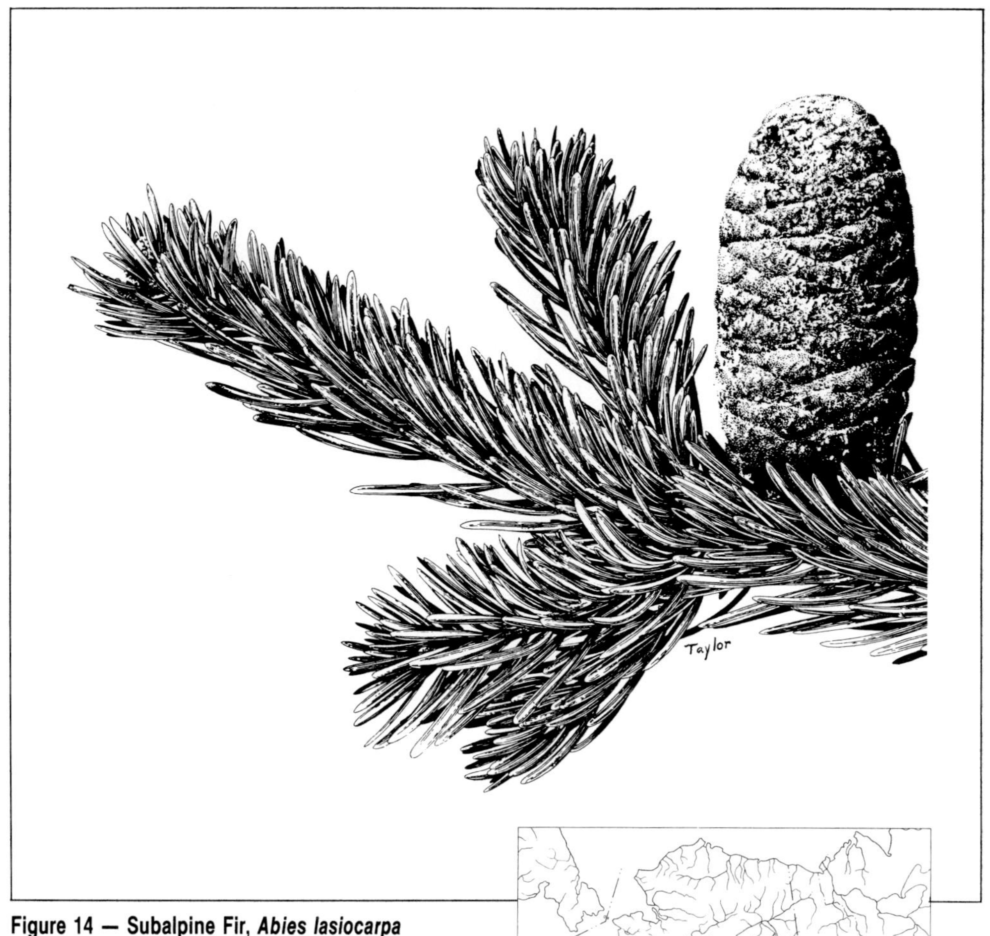

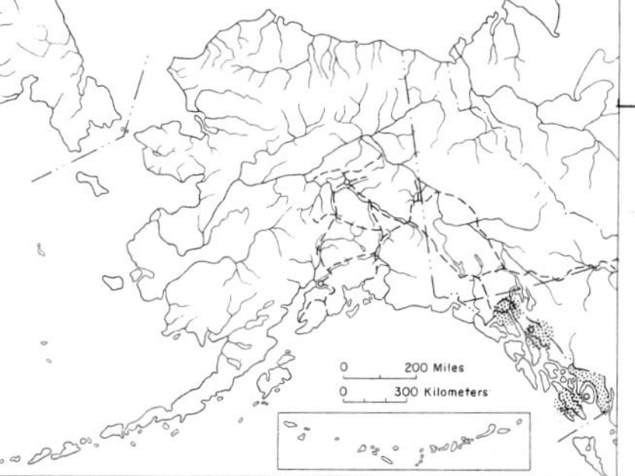

Figure 14 — Subalpine Fir, *Abies lasiocarpa*

Pacific Silver Fir (figure 15)

(Abies amabilis)

Also called: silver fir, white fir.

Pacific silver fir occurs in Alaska only at the southern tip of the Panhandle. Commonly 80 feet tall with 2-foot-diameter trunk, observers reported a specimen 149 feet tall with a 49-inch trunk.

Considered rare in Alaska, Pacific silver fir is a commercial species farther south in British Columbia. One stand in Misty Fjords National Monument east of Ketchikan has been designated a research natural area, so scientists can study this resinous and aromatic tree at the northern end of its range.

What to look for: Needles ¾ to 1¼ inches long, grooved, shiny dark green on top, silvery white on bottom, crowding the twigs. Twigs slender, gray with fine hair. Bark smooth, gray splotched with white. Cylindrical cones 4 to 5 inches long, sitting upright on top branches of tree. In autumn cones disintegrate, releasing seeds; whole cones not found on the forest floor.

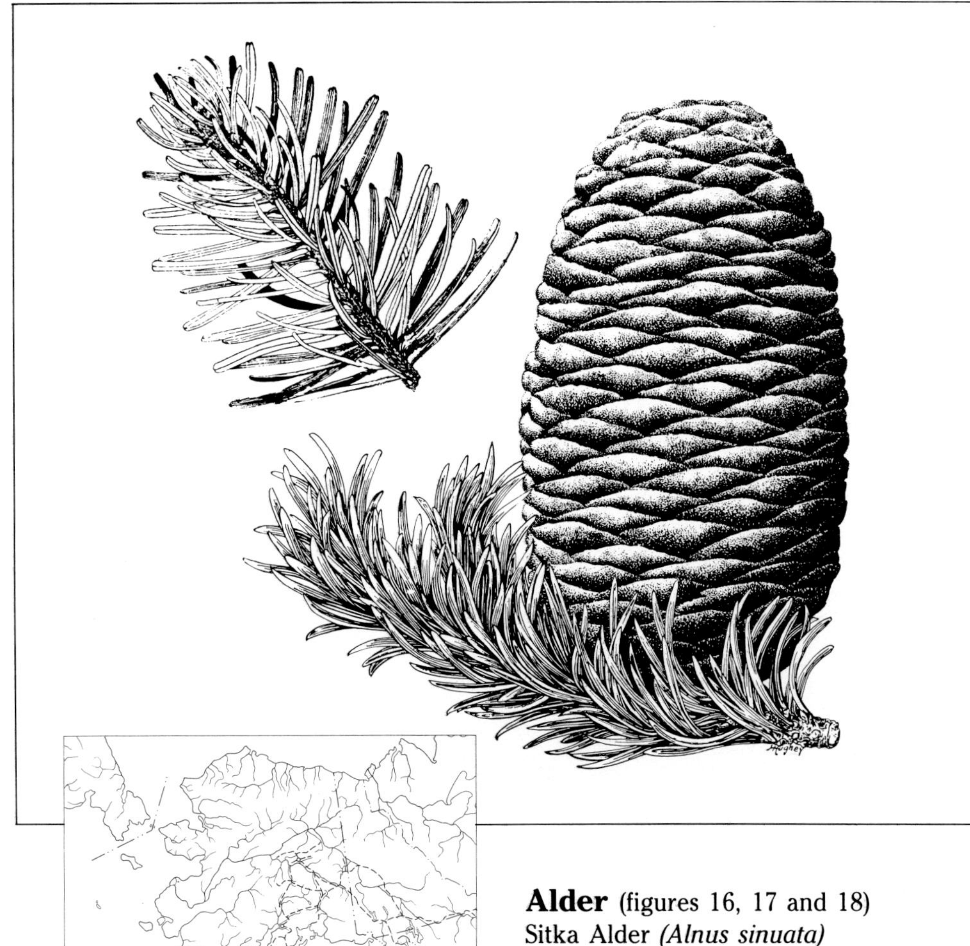

Figure 15 — Pacific Silver Fir, *Abies amabilis*

Alder (figures 16, 17 and 18)

Sitka Alder *(Alnus sinuata)*
Thinleaf Alder *(Alnus tenuifolia)*
Red Alder *(Alnus rubra)*

At least one species of alder can be found in most forested areas of Alaska. As shrubs or small trees, they are often found along beaches, streams, avalanche paths, landslides, roadsides, landings and other disturbed areas. Alders produce large numbers of tiny seeds, which can be carried many miles by the wind. They are one of the first trees to colonize sterile mineral soil. There, alders prepare the soil for subsequent plants by fixing nitrogen into the soil, adding organic matter in the form of dead wood and leaf fall, and preventing erosion.

Alders, intolerant of shade, die when conifers overtop them. With the larger red alder, this may take years, since spruce grow slowly under the alders' canopy; in some areas, foresters deliberately kill alder to allow spruce to grow faster.

Alders are cut for poles, carving wood and firewood, and are also used to smoke fish.

A feature common to all alders is the small, dried-up conelike fruit which persists on the tree long after seeds have been dispersed.

What to look for (Sitka alder): Shrub or small tree to 30 feet, with up to 8-inch-diameter trunk. Leaf blades 2½ to 5 inches long, pointed, oval, doubly toothed along leaf margins, speckled yellow-green on top, bottoms lighter and shiny, on ½- to ¾-inch stalk. Leaves sticky when young. Bark gray to light gray, smooth and thin. Male flowers on long catkins. Female flowers mature to ½- to ¾-inch-long, egg-shaped fruit on long, spreading stalks.

What to look for (thinleaf alder): Shrub or small tree to 30 feet, with up to 8-inch-diameter trunk, commonly forming clumps on trees. Leaf blades thin, blunt-pointed, oval, 1¼ to 2½ inches long, with shallow rounded lobes, doubly toothed along leaf margins, dark green on top, pale green and hairy on bottom, on ¼- to 1-inch stalks. Bark gray to

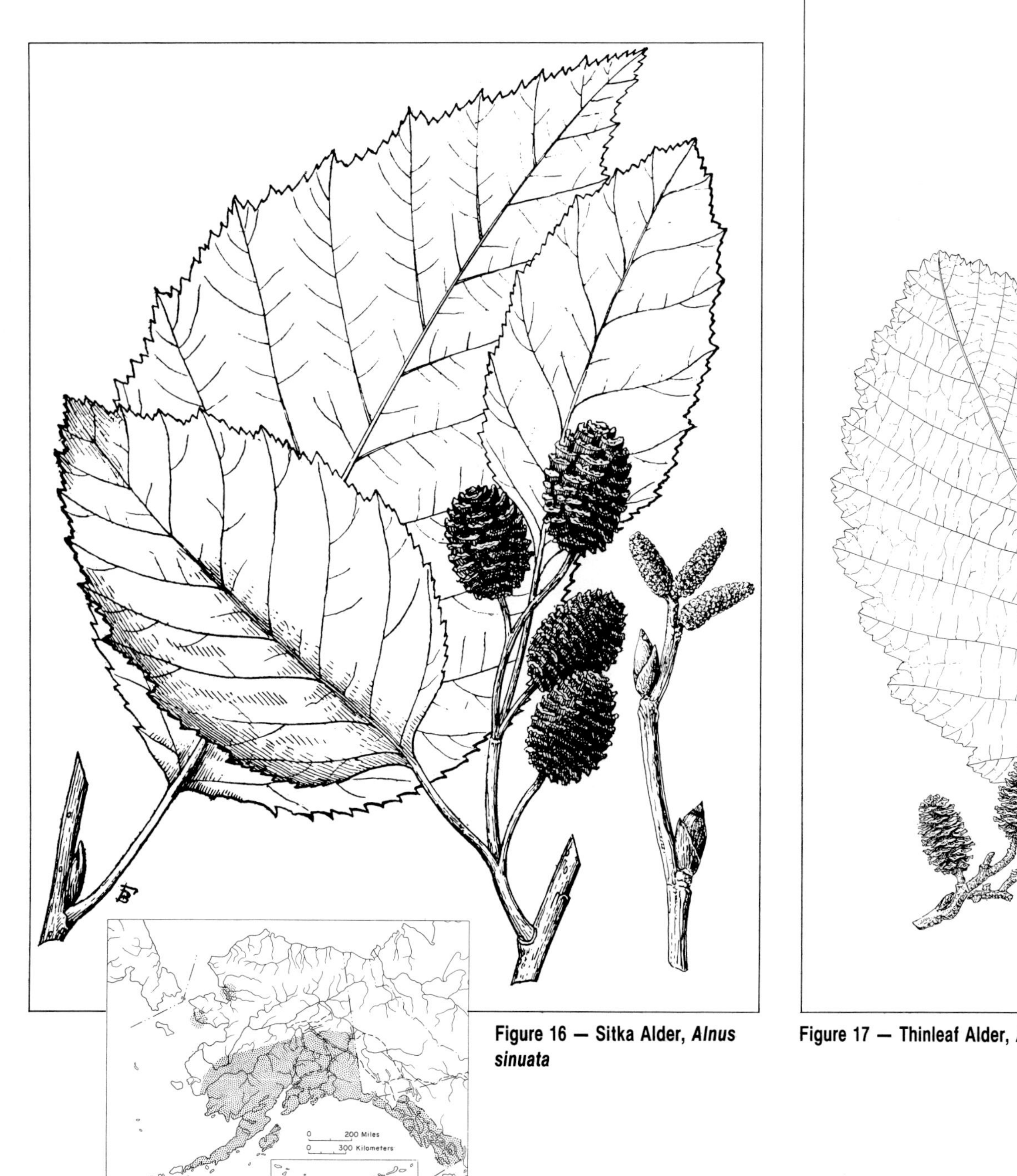

Figure 16 — Sitka Alder, *Alnus sinuata*

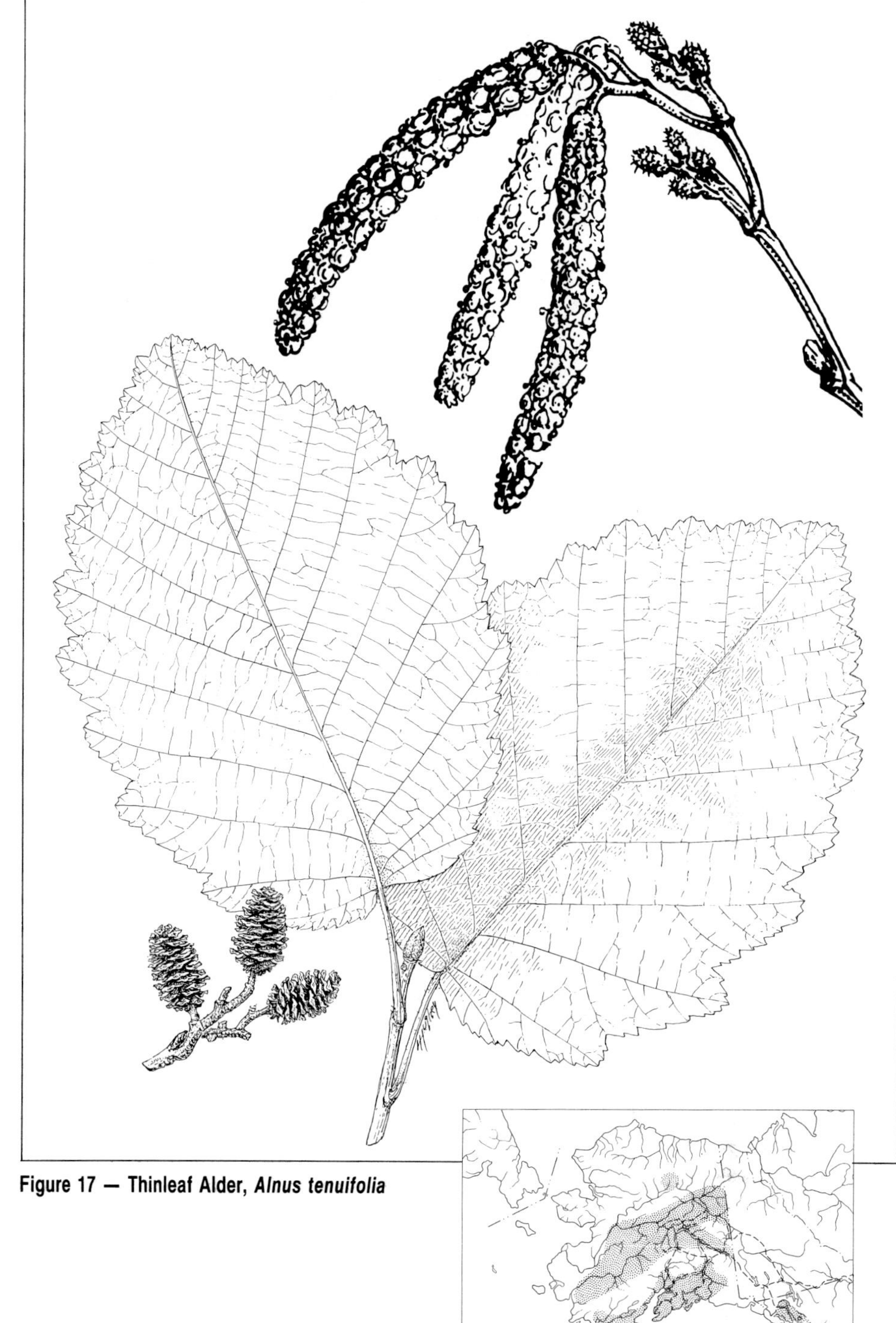

Figure 17 — Thinleaf Alder, *Alnus tenuifolia*

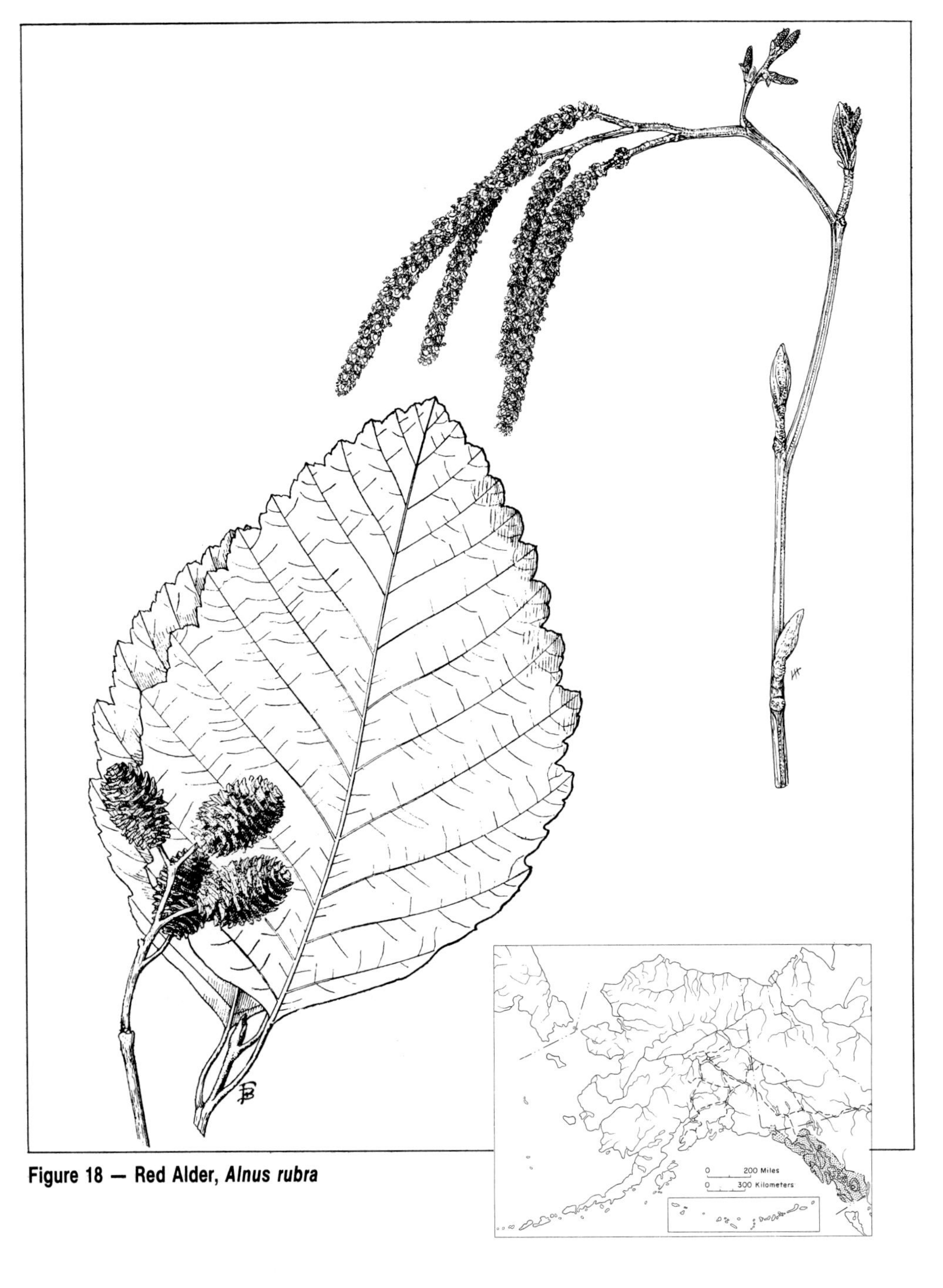

Figure 18 — Red Alder, *Alnus rubra*

dark gray, smooth, becoming reddish-gray and scaly on older trunks. Male flowers on long catkins. Female flowers mature to ⅜- to ⅝-inch egg-shaped fruits on short stalks.

What to look for (red alder): tree 20 to 40 feet tall, with straight 4- to 16-inch diameter trunk. Leaf blades 3 to 5 inches long, oval, slightly pointed at both ends, shallow lobes and doubly toothed on leaf margins. Leaves dark green on top, pale with rusty hairs along veins on bottom, on ¼- to ¾-inch stalks. Thick leaf slightly curved under at edges. Bark gray, splotched with white, smooth, becoming thin and scaly on older trunk. Male flowers on long catkins. Female flowers mature to ½- to 1-inch-long egg-shaped fruits on short stalks.

Pacific Yew (figure 19)

(Taxus brevifolia)

Also called: western yew.

This 20- to 30-foot evergreen with 2- to 12-inch conical trunk grows in the southern part of southeastern Alaska. Pacific yew is too rare in Alaska to be used commercially; farther south, its wood is made into cabinets, poles and bows.

What to look for: Needles flat, ¼ to ½ inch long, pointed but not prickly, shiny yellow on top, pale green on bottom, neatly arranged in two rows along slender twigs. Bark purplish-brown, thin, scaly, ridged and fluted. Bright red, hard heartwood; light yellow sap-

Figure 19 — Pacific Yew, ***Taxus brevifolia***

wood. Male and female flowers on different trees. Single seeds ⅜ inch long, brown, surrounded by scarlet, juicy, cuplike berry. Seeds and foliage are poisonous.

Figure 20 — Douglas Maple, ***Acer glabrum* var. *douglasii***

Douglas Maple (figure 20)

(Acer glabrum var. *douglasii)*
Also called: dwarf maple, Douglas Rocky Mountain maple.

Douglas maple, the only maple native to Alaska, thrives in southeastern Alaska along shores and tidal meadows, and occasionally on slopes. It often grows as a tall shrub, but can become a tree 20 to 30 feet tall with a 6- to 12-inch-diameter trunk.

What to look for: Three-lobed maple-style leaf blade, with irregular teeth along margins, two to four inches long, shiny dark green on top, pale with yellowish veins on bottom. Leaf stalk often as long as leaf blade. Male and female flowers grow on different trees. Small female flowers develop into typical double-winged maple seeds (**samaras**), ¾ to 1 inch long, which are red when shed in late summer.

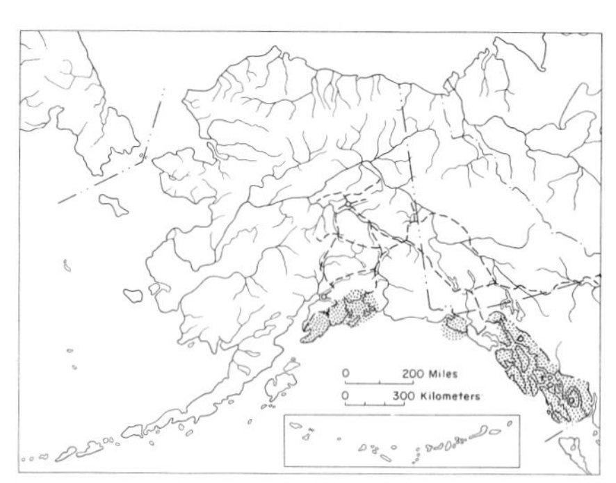

Figure 21 — Oregon Crab Apple, *Malus diversifolia*

Oregon Crab Apple (figure 21)

(Malus diversifolia)

Also called: western crab apple, wild crab apple.

Commonly a small shrub, Oregon crab apple also grows to 25 feet with trunk diameters to 5 inches. Found in thickets, or as scattered individual trees in beach meadows, along muskegs, river bottoms and other lowlands in southeastern Alaska, on the Kenai Peninsula and along Prince William Sound. Wood is used for tool handles and smoking fish. Small, sour apples can be eaten raw, or made into jams or jellies.

What to look for: Leaf blades elliptical, 1½ to 4 inches long, sharply pointed teeth along margins, shiny green on top, pale and often hairy on bottom, attached to twig by slender 1- to 2-inch stalks. White or pink flowers with five petals blossom in June; ½- to ¾-inch, thin-skinned, red or yellow fruit matures from August to October.

Pacific Red Elder (figure 22)

(Sambucus callicarpa)

Also called: elderberry, stinking elder, scarlet elder, redberry elder.

This shrub occasionally attains the stature of a small tree; a stand of red elder in southeastern Alaska was reported to be 20 feet tall with 5-inch-diameter trunks. Red elder is common in moist locations along the coast of Alaska from Ketchikan to the tip of the Alaska Peninsula. Berries are not considered edible while raw, but are made into jellies and wine; the berries' seeds are poisonous.

What to look for: Plant releases strong odor when leaves or stems are crushed. Compound leaves contain five to seven short-stalked leaflets, 2 to 5 inches long, which are

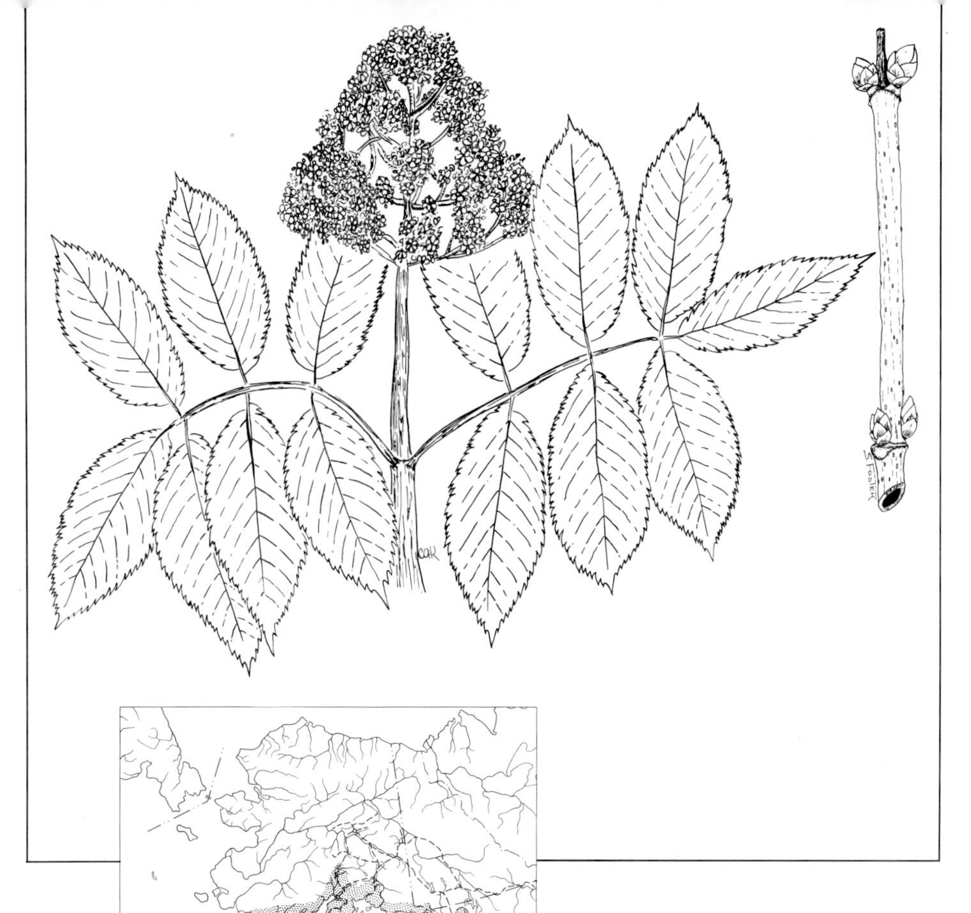

Figure 22 — Pacific Red Elder, *Sambucus callicarpa*

arranged in opposite pairs except for odd leaflet at the tip. Tops of leaflets are green and smooth; bottoms paler and hairy. Twigs stout, gray with brown raised dots. Bark light to dark gray or brown, smooth on younger limbs, cracked and furrowed on older growth. Flower clusters, 2 to 4 inches long, erect, with many tiny white flowers, blossoming in May and June. Bright red or orange, 3/16-inch, round berries mature in July and August.

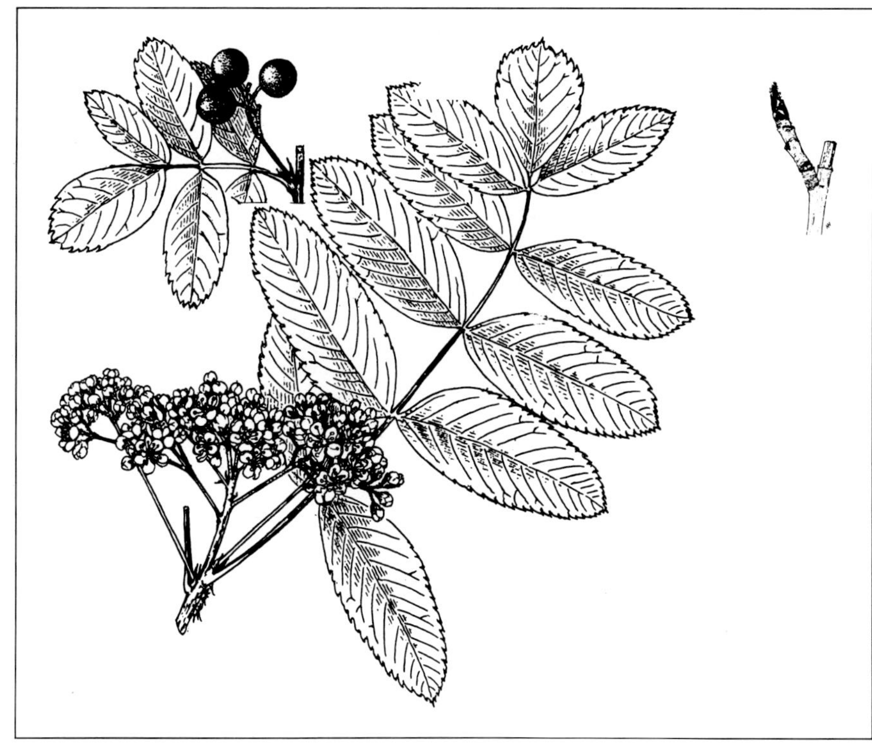

Sitka Mountain-Ash (figure 23)

(Sorbus sitchensis)
Also called: western mountain-ash, Pacific mountain-ash.

Sitka mountain-ash grows as a shrub or small tree 15 to 20 feet tall with trunk up to 6 inches in diameter. Uncommon or rare in coastal forests from the Panhandle to the Alaska Peninsula.

What to look for: Compound leaves of 7 to 13 leaflets, arranged in pairs except for odd leaflet at tip. Leaflets 1¼ to 2½ inches long, elliptical, rounded, margins toothed above middle, dull blue-green on top, pale on bottom. Twigs stout, rusty and hairy when young, becoming gray in older growth. Bark smooth, gray. Flower clusters round, 2 to 4 inches across, containing 15 to 60 tiny flowers which blossom in June. Red fruit, 3/8 to ½ inch long, becomes orange and purple with age; fruit matures in August and September.

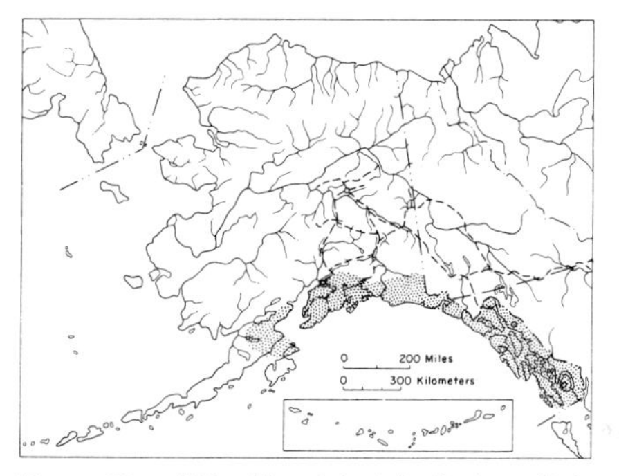

Figure 23 — Sitka Mountain-Ash, *Sorbus sitchensis*

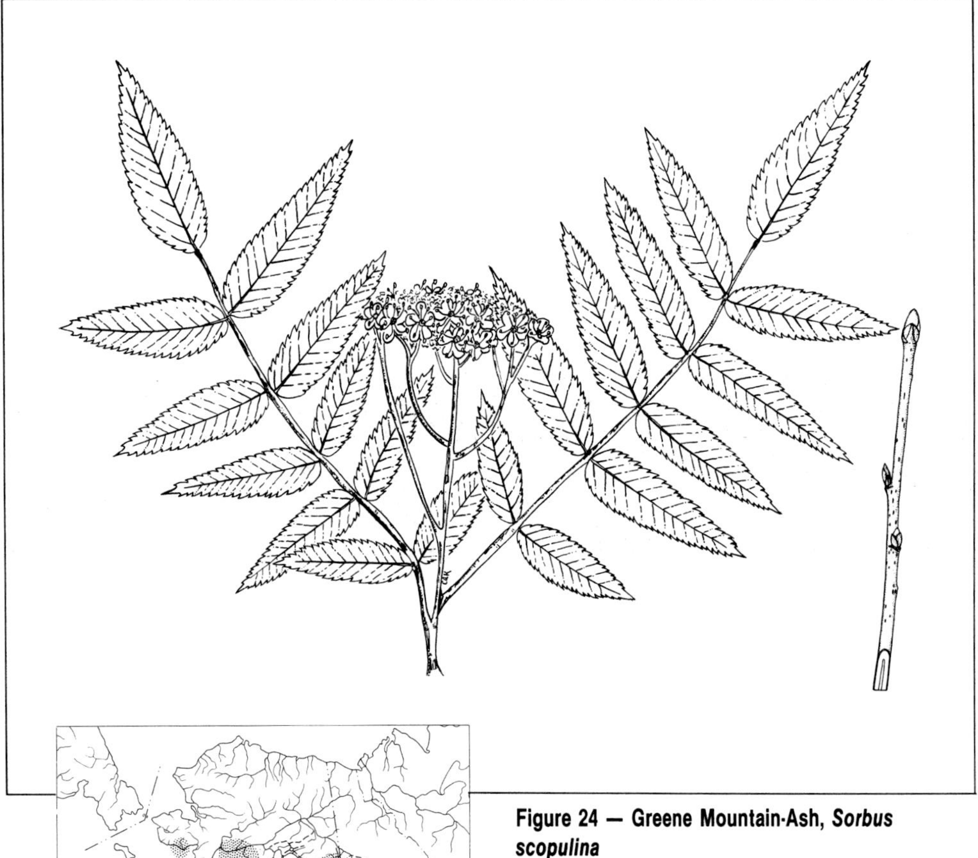

Figure 24 — Greene Mountain-Ash, *Sorbus scopulina*

Greene Mountain-Ash (figure 24)

(Sorbus scopulina)

Also called: western mountain-ash.

Usually a shrub, Greene mountain-ash has been recorded as a 20-foot tree near Haines. It grows in openings and clearings across southcentral Alaska to Canada, and on the mainland of southeastern Alaska.

What to look for: Compound leaves containing 11 to 15 dagger-shaped leaflets, paired except for the odd leaflet at tip. Leaflets 1¼ to 2½ inches long, margins sharply toothed, shiny green tops, paler bottoms. Twigs have scattered elliptical dots. Bark smooth, gray. Flower clusters 1¼ to 3 inches across, bearing many tiny flowers, blossom in June and July. Shiny, ⅜-inch, round red fruits mature in July, with many remaining on the tree through the winter.

Pacific Serviceberry (figure 25)

(Amelanchier florida)

Also called: western serviceberry, juneberry.

Pacific serviceberry grows as a shrub or tree to 16 feet with a 5-inch-diameter trunk. This species of serviceberry grows in forests and openings in southeastern Alaska, the Cook Inlet area and parts of the Alaska Peninsula. The sweet fruits are eaten raw, prepared in biscuits and puddings, made into jam, or dried like raisins.

What to look for: Leaf blades rounded, elliptical, 1 to 2 inches long, coarsely toothed, dark green on top, paler on bottom, on slender stalks. Young twigs reddish-brown. Bark gray or brown, thin, smooth or slightly fissured. Flower clusters erect, 1½ to 3 inches long, with several fragrant 1-inch white flowers which blossom in June and July. Rounded purple fruit, ⅜ to ½ inch long, juicy, containing a few brown seeds, maturing in August and September.

Willow

Eight species of willow in Alaska can grow to tree size. Positive identification is often difficult, since distinctions between species can be subtle. Features also may vary depending on location, and some species hybridize where their ranges overlap.

Willows are **dioecious**, having male and female catkins on different plants. After fer-

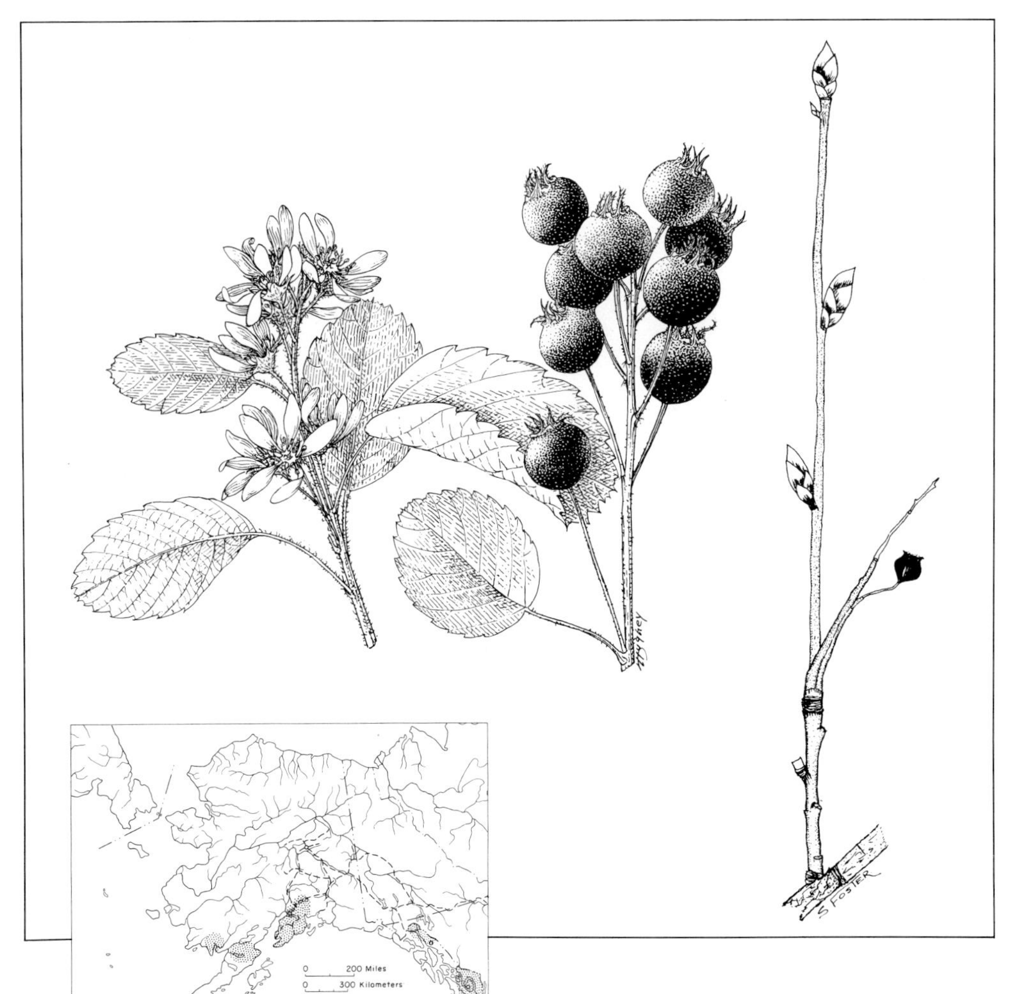

Figure 25 — Pacific Serviceberry, *Amelanchier florida*

tilization, female flowers develop into seed capsules, which split open to release numerous tiny airborne seeds with cottony tufts. During winter, willow buds are covered by a single bud scale. Willow bark has a bitter, aspirinlike taste.

Littletree Willow (figure 26)
(Salix arbusculoides)

One of the most common willows in Alaska, littletree willow grows in dense thickets along waterways throughout the Interior. This species can become a 25- to 30-foot tree, with 5- to 6-inch trunk diameter. Forms diamond willow (see Bebb willow).

What to look for: Leaves narrow, dagger-like, 1 to 3 inches long, finely toothed edges,

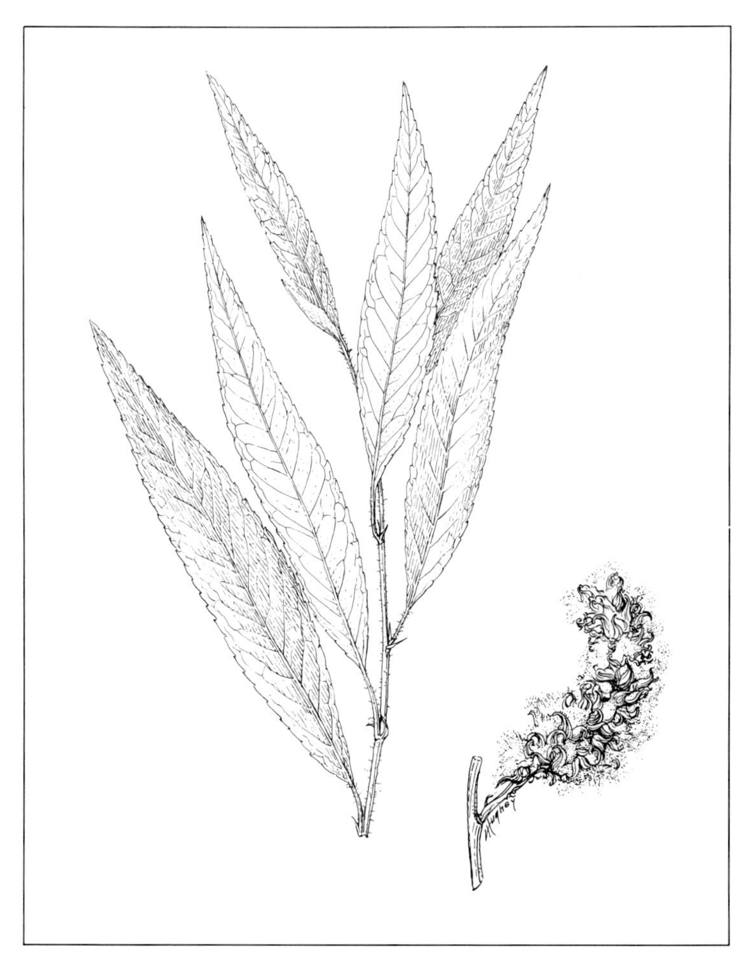

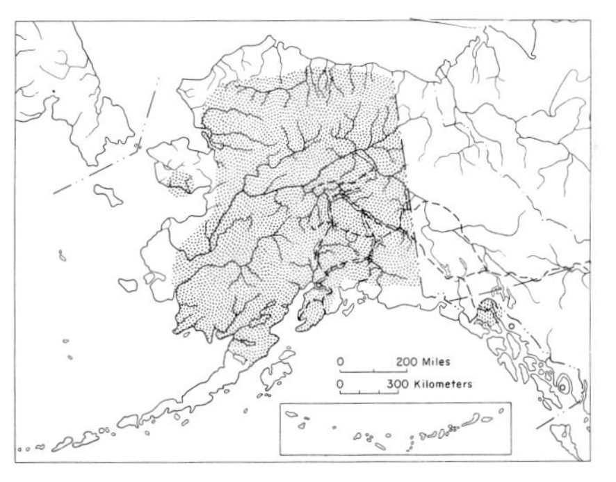

Figure 26 — Littletree Willow, *Salix arbusculoides*

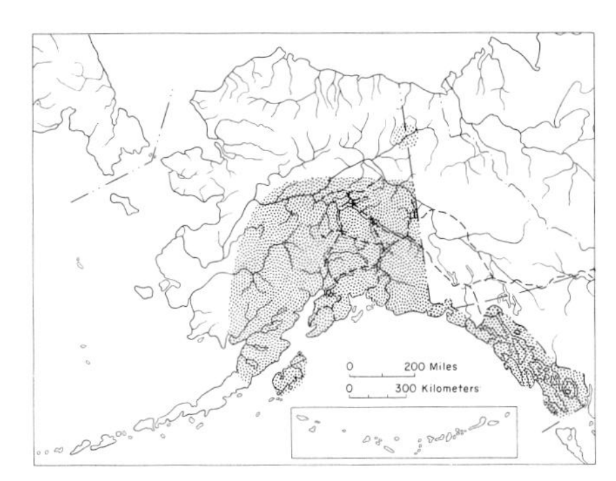

Figure 27 — Scouler Willow, *Salix scouleriana*

green on top, bottoms whitish with silver hair. Twigs slender, branched, yellow-brown when young, red-brown and shiny in older growth. Bark gray to red-brown. Catkins 1 to 2 inches long, flowering in May and June.

Scouler Willow (figure 27)

(Salix scouleriana)
Also called: mountain willow, black willow, fire willow.

Scouler willow, the most common willow in southeastern and southcentral Alaska, is also called fire willow because it rapidly revegetates burned areas. It is an important moose **browse**; many trees are stripped of bark by moose. Forms diamond willow (see Bebb willow).

What to look for: Leaves resemble tongues, 2 to 5 inches long, tapering to a narrow base, dark green and hairless on top, gray to rust-colored hair on bottom. Twigs stout, yellow- or green-brown and hairy when young; dark brown and nearly hairless on older growth. Bark smooth, thin, gray, becoming dark brown and ridgelike in older growth. Numerous stout catkins, 1 to 2 inches long, flower in May. Gray, woolly seed capsules release seeds in June.

Figure 28 — Bebb Willow, *Salix bebbiana*

Bebb Willow (figure 28)

(Salix bebbiana)

Also called: diamond willow, beak willow.

Common on uplands of interior Alaska, Bebb willow contributes significantly to the diet of moose and other animals. This species can be a bushy tree 15 to 35 feet tall, with trunk 6 to 9 inches in diameter.

Diamond willow comes most frequently from this species; fungi form cavities at junctions of branches and the main stem which create diamond-shaped patterns. Woodcarvers exploit these patterns to create unique canes, furniture, lampposts and even parts of buildings.

What to look for: Leaves elliptical, 1 to 3½ inches long, without teeth on margins, dull green on top, gray or whitish on the bottom,

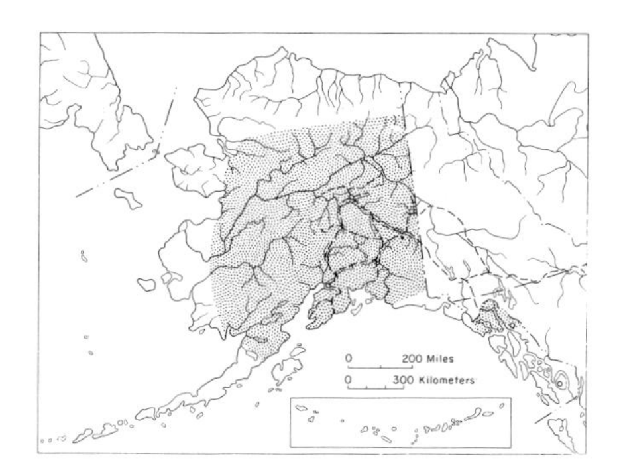

hair on both sides decreasing with age. Twigs are slender, yellowish to brown, hairy when young. Bark gray to dark gray, smooth, becoming rough and furrowed on older growth. Erect 1- to 3-inch-long catkins, flowering in May and June. Relatively long, slender seed capsules, shedding by mid-July.

Feltleaf Willow (figure 29)

(Salix alaxensis)

This species grows throughout Alaska, from coastal islands to the Arctic Slope. The tree form of this willow reaches 20 to 30 feet tall, with trunks 4 to 7 inches in diameter. Moose in search of browse have been known to pull down larger branches; usually the tree outgrows the animals. People, too, reportedly eat the inner bark of feltleaf willow. In some parts of Alaska it serves as the only firewood source.

Trunks sometimes have diamond willow patterns where twigs have died [see Bebb willow].

What to look for: Leaves tongue-shaped, 2 to 4 inches long, dull green on top, dense hair on bottom; hence the name. Older twigs stout

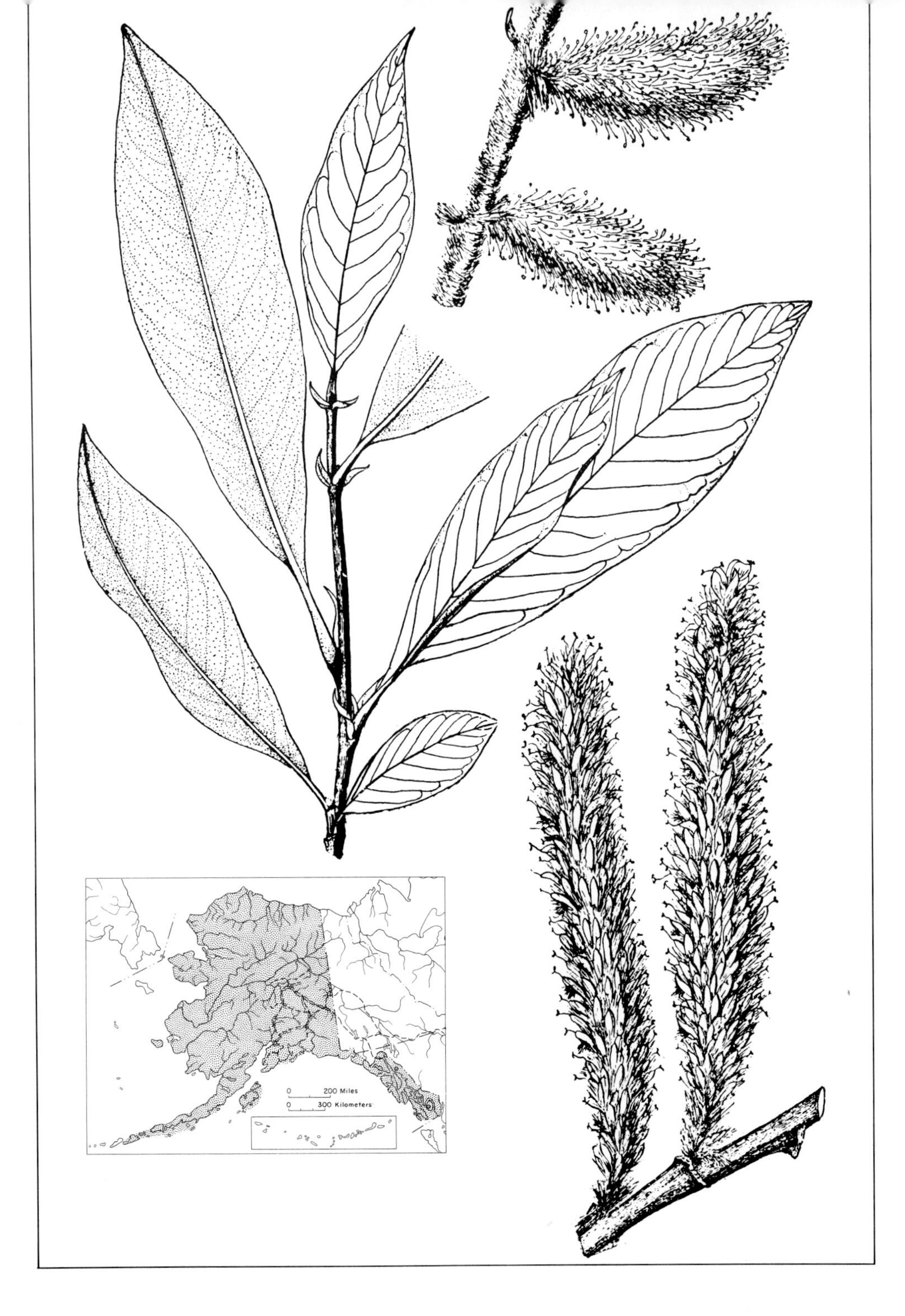

Figure 29 — Feltleaf Willow, *Salix alaxensis*

with white wool. Bark gray and smooth, becoming rough and furrowed in older growth. Catkins 2 to 4 inches long, with blackish scales, appear before the leaves. Relatively long, white, woolly seed capsules; seeds ripen in June and July.

Sitka Willow (figure 30)
(Salix sitchensis)
Also called: silky willow.

Found in southcentral and southeastern Alaska, this species can grow into trees 10 to 20 feet tall with 4- to 6-inch-diameter trunks. Wood can be used for smoking fish.

What to look for: Leaves tongue-shaped, 2 to 4 inches long, dark green on top, bottoms with silvery, silky hairs that give them a distinctive sheen. Twigs slender, may have fine hair when young, becoming reddish-brown and hairless with age. Bark gray, smooth, becoming scaly and slightly furrowed in older growth. Slender catkins, 2 to 4 inches long, appear with leaves, flower in May. Short silver-haired seed capsules, shedding by July or early August.

Grayleaf Willow (figure 31)
(Salix glauca)

Generally a shrub, grayleaf willow is occasionally found as a tree up to 20 feet tall with a 5-inch-diameter trunk. Grows throughout Alaska except for the Aleutians and parts of

Figure 30 — Sitka Willow, *Salix sitchensis*

Figure 31 — Grayleaf Willow, *Salix glauca*

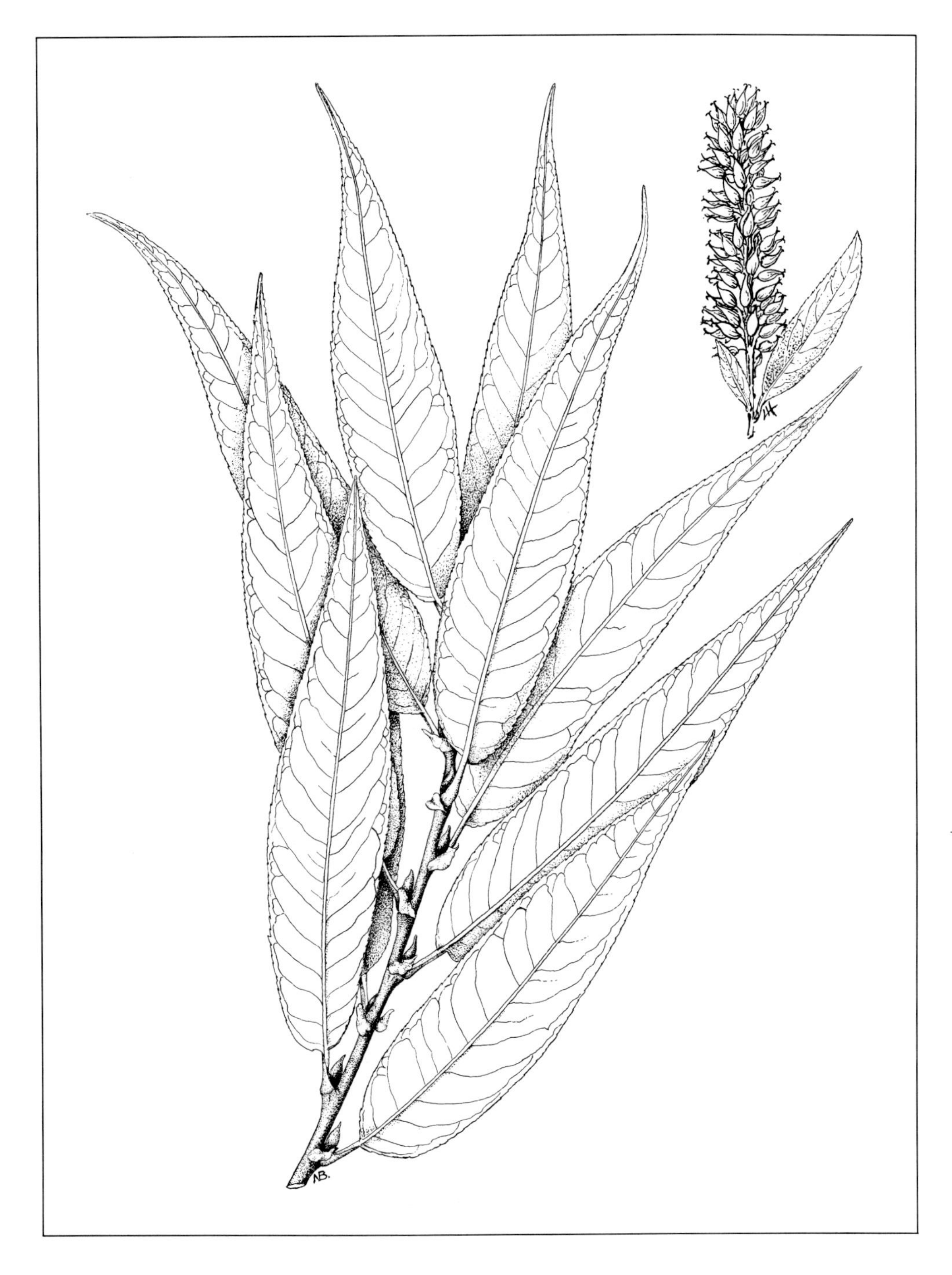

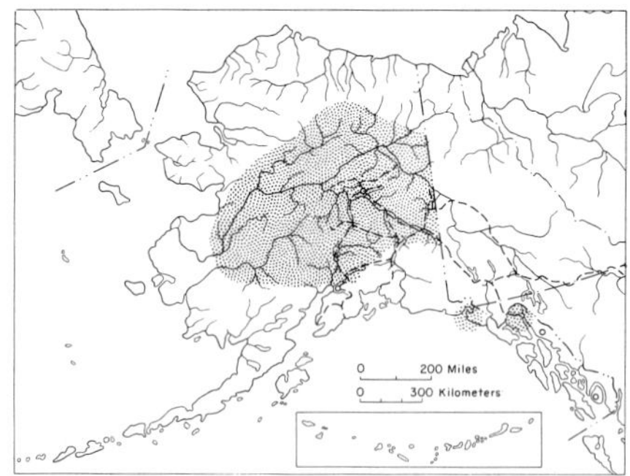

Figure 32 — Pacific Willow, *Salix lasiandra*

the Panhandle. Grayleaf willow is a common pioneer species after fires and other disturbances.

What to look for: Many variations in this species make identification difficult. Overall appearance is gray. Leaves in oval to dagger-like shapes, 1½ to 3½ inches long, short-pointed or rounded at the tip, with occasional sticky teeth on lower edges. Tops of leaves green (sometimes hairy); bottoms whitish with scattered hairs. Young twigs have dense white hairs; older twigs gray, hairy or hairless. Bark gray, smooth, becoming rough and furrowed on older growth. Catkins ¾ to 2 inches long, with light brown to yellow scales, hairy, flowering in June. Hairy seed capsules; seeds ripen in July and August. Catkins sometimes remain on plant well into winter.

Pacific Willow (figure 32)

(Salix lasiandra)

Also called: western black willow, yellow willow.

Shrub or tree to 20 feet tall, Pacific willow is found in Interior and northern southeastern

Alaska. Grows with other willows along rivers, and occasionally on upland sites. Farther south in the Lower 48, this tree can attain heights to 60 feet.

What to look for: Leaves lance-shaped, pointed, 2 to 5 inches long, finely toothed margins, shiny green on top, whitish and hairy on bottom. Twigs stout, hairy when young, becoming reddish and shiny with age. Bark gray, smooth, becoming rough and deeply furrowed in older growth. Catkins 2 to 4 inches long, appear with the leaves. Hairless seed capsules.

Hooker Willow (figure 33)
(Salix hookeriana)
Also called: bigleaf willow, Yakutat willow.

Hooker willow grows as a shrub or tree commonly 10 to 16 feet tall (occasionally to 25 feet), with 8- to 15-inch trunk diameter. Local to the Yakutat area, rare in other southcentral coast sites. Browsed by moose.

What to look for: Leaves oval, 1½ to 3 inches long, edges toothless or slightly wavy, pale green on top, whitish on bottom. Leaves hairy while unfolding, losing hair with age. Twigs stout, dark brown, densely covered with gray or white wool for two to three years. Smooth gray bark. Catkins 3 to 4 inches long, covered with long whitish hairs. Flowers in May and June; seeds ripen in June and July.

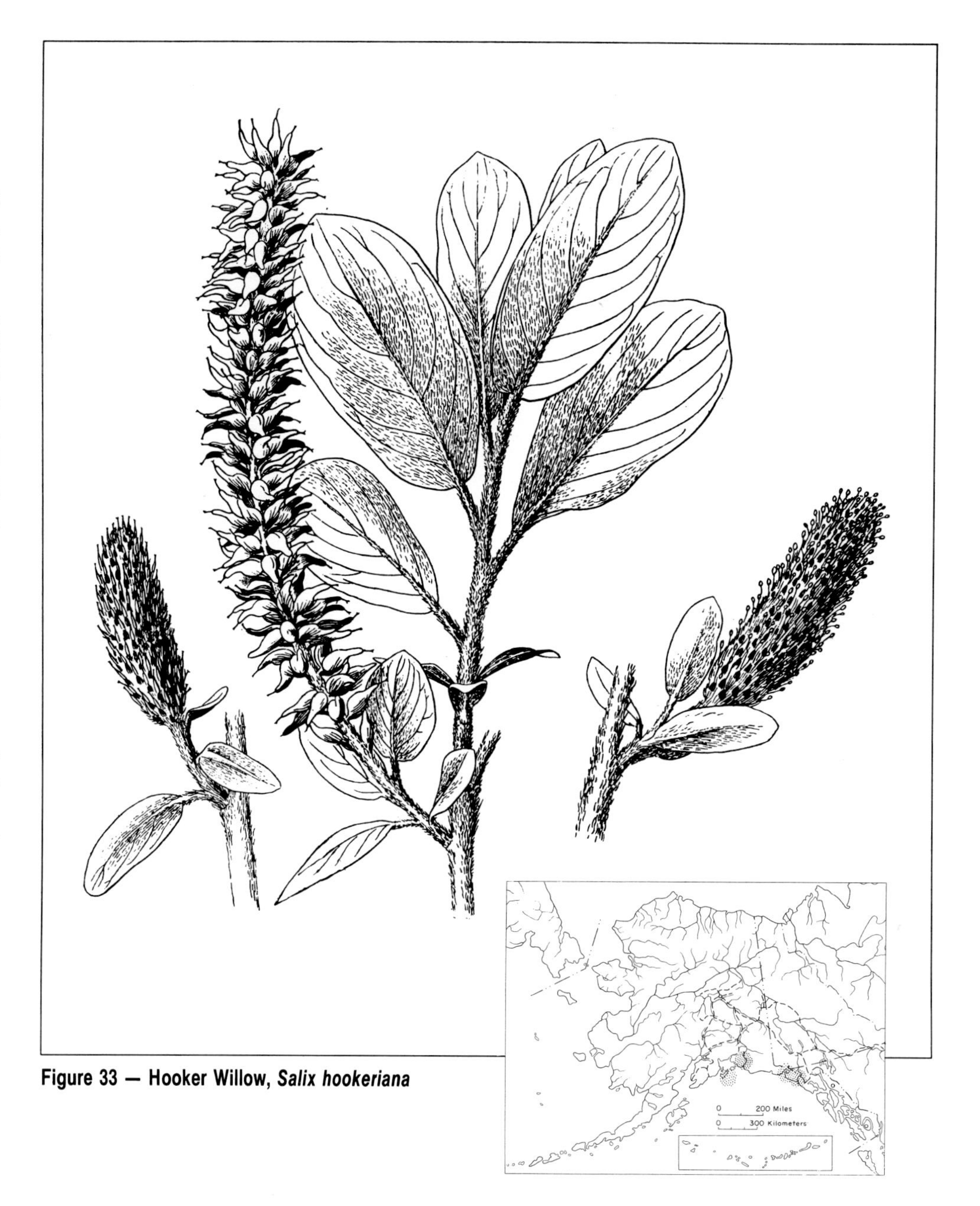

Figure 33 — Hooker Willow, *Salix hookeriana*

Measuring Alaska's Forests

In 1856 A.R. Roche, a visitor to Russian America, recorded that "A great portion of this vast region . . . is covered with forests of the largest and most valuable trees." Writing just before the sale of Alaska to the United States in 1867, S.N. Buynitzky noted that the Interior consisted of "primeval forests of such thickness that the only ways of communication are rivers."

Roche and Buynitzky obviously didn't have extensive firsthand experience in Alaska, particularly in the Interior. They echoed the common misconception that Alaska was a vast storehouse of timber — at least the areas that weren't covered by perpetual ice.

Scientific measurements of Alaska's forests didn't start until after 1900, when the federal government realized it had better learn more about the frontier that was suddenly attracting thousands of fortune seekers. Trained foresters visited Alaska; their reports dispelled the myths, but were not inventories. Reliable accounting of Alaska's forests wasn't possible until accurate mapping and aerial photography were developed after World War II.

Forests in a Nutshell

About one-third of Alaska's land area is considered to be forested. Less than one-fourth of this area is classified as **commercial forestland**, capable of growing at least 20 cubic feet of wood per acre each year. Alaska has about the same amount of commercial forestland as Oregon.

Students in a workshop given by the Alaska Division of Forestry learn to use instruments that help them measure stands of trees more quickly and accurately.
(Tony Gasbarro, Cooperative Extenstion Service)

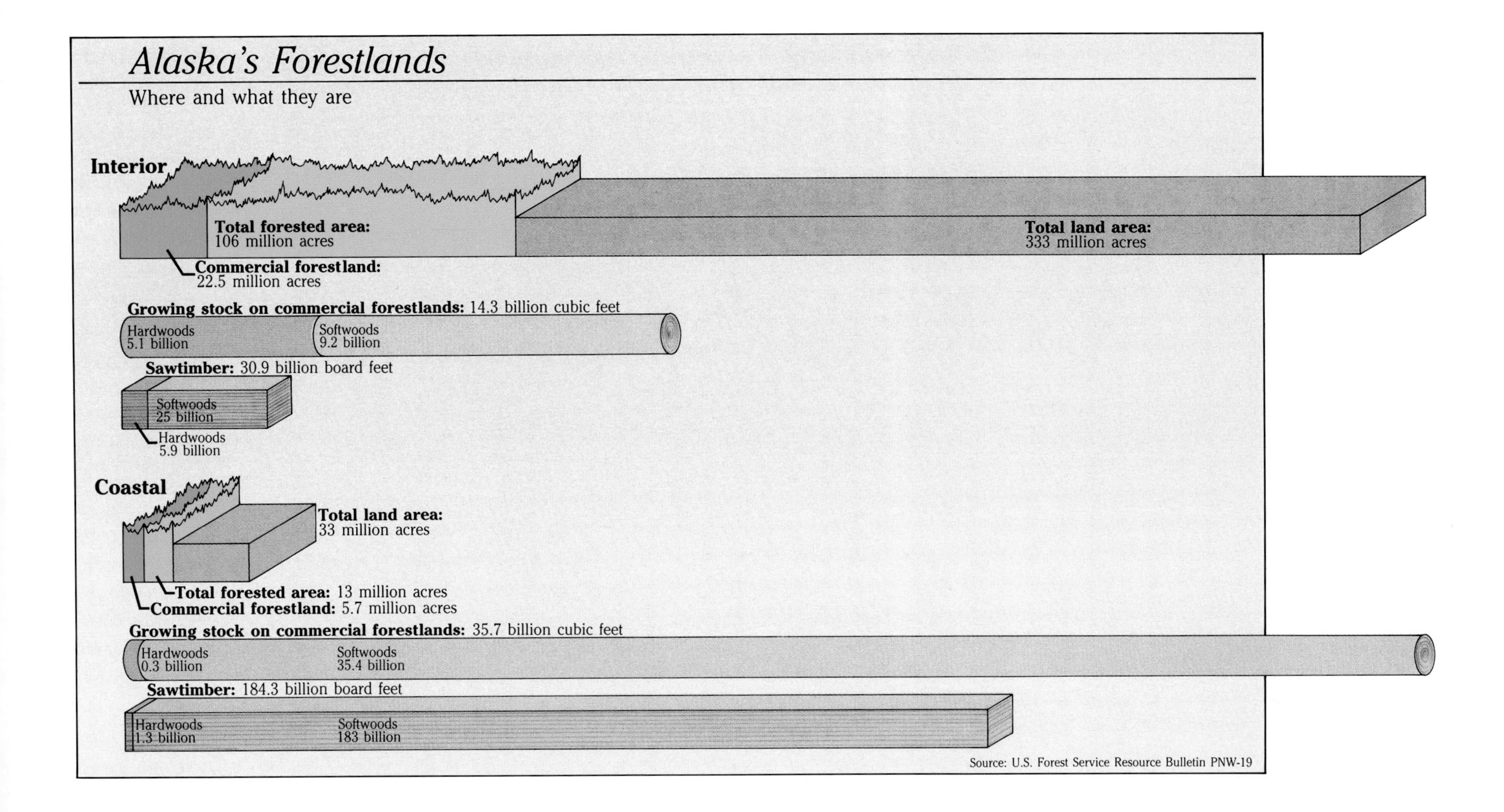

Alaska's Forestlands
Where and what they are
Interior
Total forested area:
106 million acres
Total land area:
333 million acres
Commercial forestland:
22.5 million acres
Growing stock on commercial forestlands: 14.3 billion cubic feet
Hardwoods
5.1 billion
Softwoods
9.2 billion
Sawtimber: 30.9 billion board feet
Softwoods
25 billion
Hardwoods
5.9 billion
Coastal
Total land area:
33 million acres
Total forested area: 13 million acres
Commercial forestland: 5.7 million acres
Growing stock on commercial forestlands: 35.7 billion cubic feet
Hardwoods
0.3 billion
Softwoods
35.4 billion
Sawtimber: 184.3 billion board feet
Hardwoods
1.3 billion
Softwoods
183 billion
Source: U.S. Forest Service Resource Bulletin PNW-19

As might be expected, interior Alaska — 10 times larger than the coastal region — has more commercial forestland. But the amount of live wood, the **growing stock volume**, in interior forests is considerably smaller. Furthermore, coastal trees generally grow to larger sizes, so the volume of **sawtimber** — live trees of a size suitable for lumber — is much larger on the coast.

Growth Rates

In southeastern Alaska, old growth stands (in which most trees are more than 150 years old), account for 90 percent of the total volume of commercial forestlands. Growth in these stands is slow, offset by disease, decay and death. Near Juneau, old growth stands increase their volume by 0.62 percent each year, but lose 0.34 percent to mortality; their **net annual growth** is therefore only 0.28 percent. Around Ketchikan, the natural mortality of sawtimber in old growth stands exceeds growth: on a cubic foot basis, their net growth rate is minus 0.16 percent.

Younger stands have faster rates of growth, approaching 2 percent per year in unmanaged stands. Where land managers wish to emphasize timber production, they seek to replace old growth with **young growth**. To further increase timber productivity, they might manage the **new growth** intensively to boost growth beyond the natural rates. [See "Growing Trees," page 68.]

In the Interior, the natural net growth on

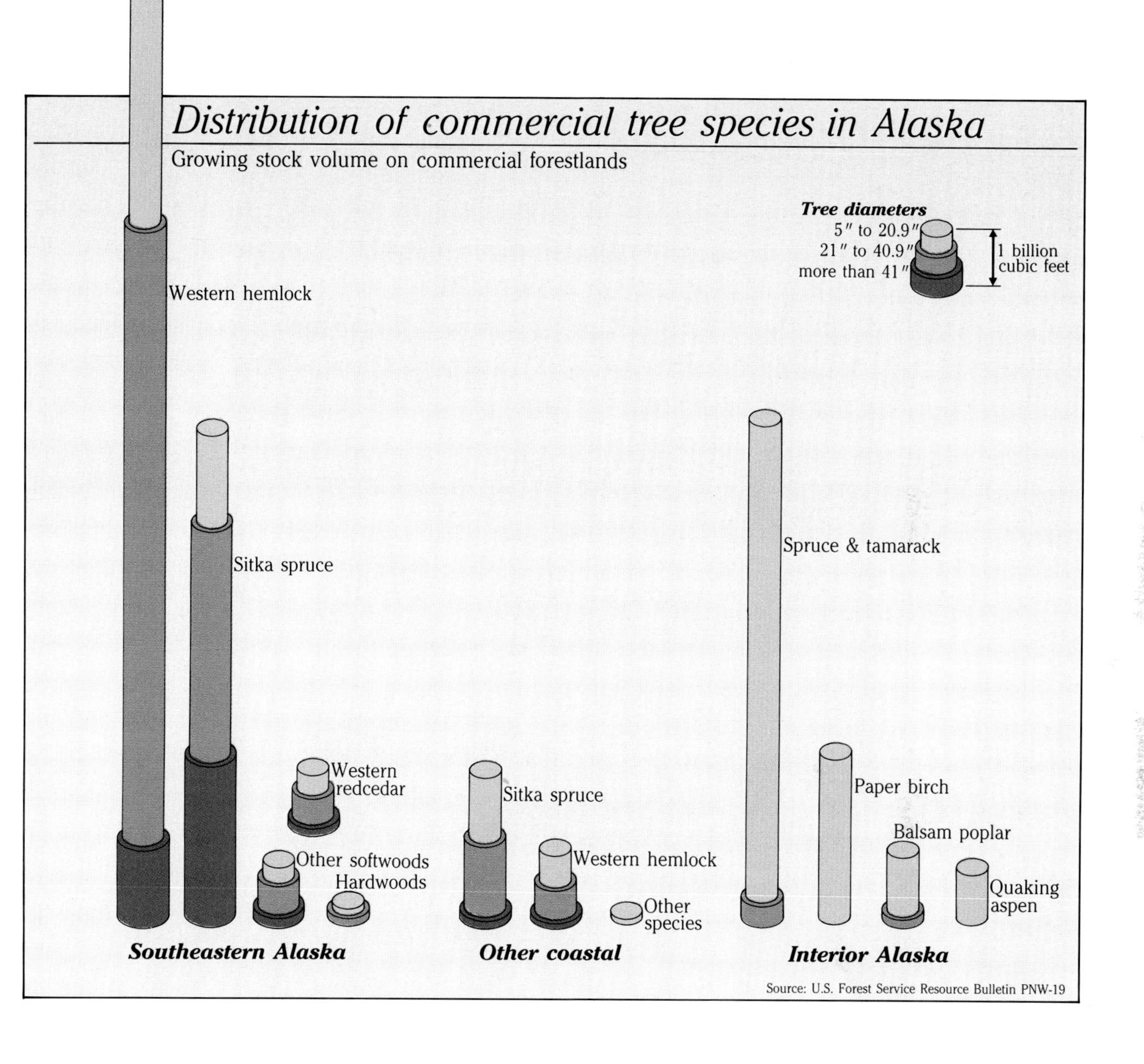

Measuring Trees

Lumber is sold by the **board foot**, a unit equal to a one-inch-thick board measuring 12 inches by 12 inches. When a raft or deck of logs is scaled, the scaler estimates how many board feet can be sawed from the logs. There are many methods of estimating; the most common in Alaska are the Scribner Rule and the International ¼-Inch Rule. Both are sets of instructions and tables enabling trained people to get consistent estimates for the same logs.

In forest inventory or research work, wood volumes are usually expressed in cubic feet or cubic meters. Like their log-scaling counterparts, timber cruisers use consistent measurements to derive volume estimates of standing trees.

One might think that a cubic foot is equal to 12 board feet, such as a stack of 12 1″ x 12″ x 12″ boards. Not so. Tables for estimating board feet in a log make allowances for the trimming necessary to cut a round piece of wood into squared boards and the volume of sawdust created by sawing. Really small trees cannot be made into lumber at all, so they have zero volume on a board-foot basis. Depending on the size of larger trees, there can be anywhere from 5 to 9 board feet in a cubic foot.

Since writing zeros is tedious, abbreviations are used to designate large quantities: **MBF** is 1,000 board feet; **MMBF** is 1,000,000 board feet.

A forester uses an instrument to determine the height of a tree, one of several measurements necessary to estimate wood volume.
(Forestry Sciences Lab)

Annual rings of this white spruce from the Copper River Valley show excellent growth.
(Forestry Sciences Lab)

commercial forestlands averages about 1.5 percent per year. Although white spruce is the primary commercial species, highest growth rates are achieved by balsam poplar, black cottonwood and hybrids of the two. In natural, unmanaged stands on good sites, annual growth of these hardwoods may approach 100 cubic feet per acre.

Forest inventories in Alaska have used national standards based on production of "industrial wood" from larger trees, so results would fit national statistical categories. Foresters in the Interior feel that attention should also be given to forests with annual growth of less than 20 cubic feet per acre. They note that Scandinavians consider lands producing 15 cubic feet per acre per year as commercial; if this same standard was applied to Alaska's Interior, the area considered commercial forest would increase by about 20 percent.

Continuing Inventory

The forest statistics reported above are from the only scientific, statewide inventory taken of Alaska's forests to date. Completed by the U.S. Forest Service in the mid-1960s, the survey involved examining thousands of aerial photographs, and field-checking enough of them to get statistically valid estimates. Using instruments that give the photographs a third dimension, trained interpreters determined species, tree heights, volume and many other characteristics of the photographed forest.

More detailed inventories of various areas of the state are continuing. Completed so far are units along the Tanana, Koyukuk, Kantishna, Copper and Kuskokwim rivers, the Susitna Valley, part of western Cook Inlet and areas of southeastern Alaska.

The Forest Service is also coordinating a reinventory of Alaska's forests, area by area, compiling not only timber data but also information about soils, other vegetation and wildlife habitat. This reinventory will use Landsat satellite scannings, high and low altitude infrared aerial photography and on-ground sampling. The total cost of this inventory is estimated at less than two cents per acre. But due to budget restraints, the work may not be finished for 20 years.

It is not necessary to cut down a tree to count its growth rings. Using an increment corer, Dennis Waterhouse extracts a cross section (example shown in inset) of a tree on Zarembo Island.
(Linda Brownstein, USFS; inset, Richard Tindall)

This Landsat high-altitude scanning of the central Susitna Valley covers 13,000 square miles. Rock, soil, water, snow, forest and other ground covers reflect different wavelengths of light. Landsat samples these reflections, records the information in a myriad of dots (called pixels) and makes up an image. In this standard false-color image, vegetation shows up as red and orange, and bare earth, water or ice appear blue or white.
(Courtesy of Forestry Sciences Lab)

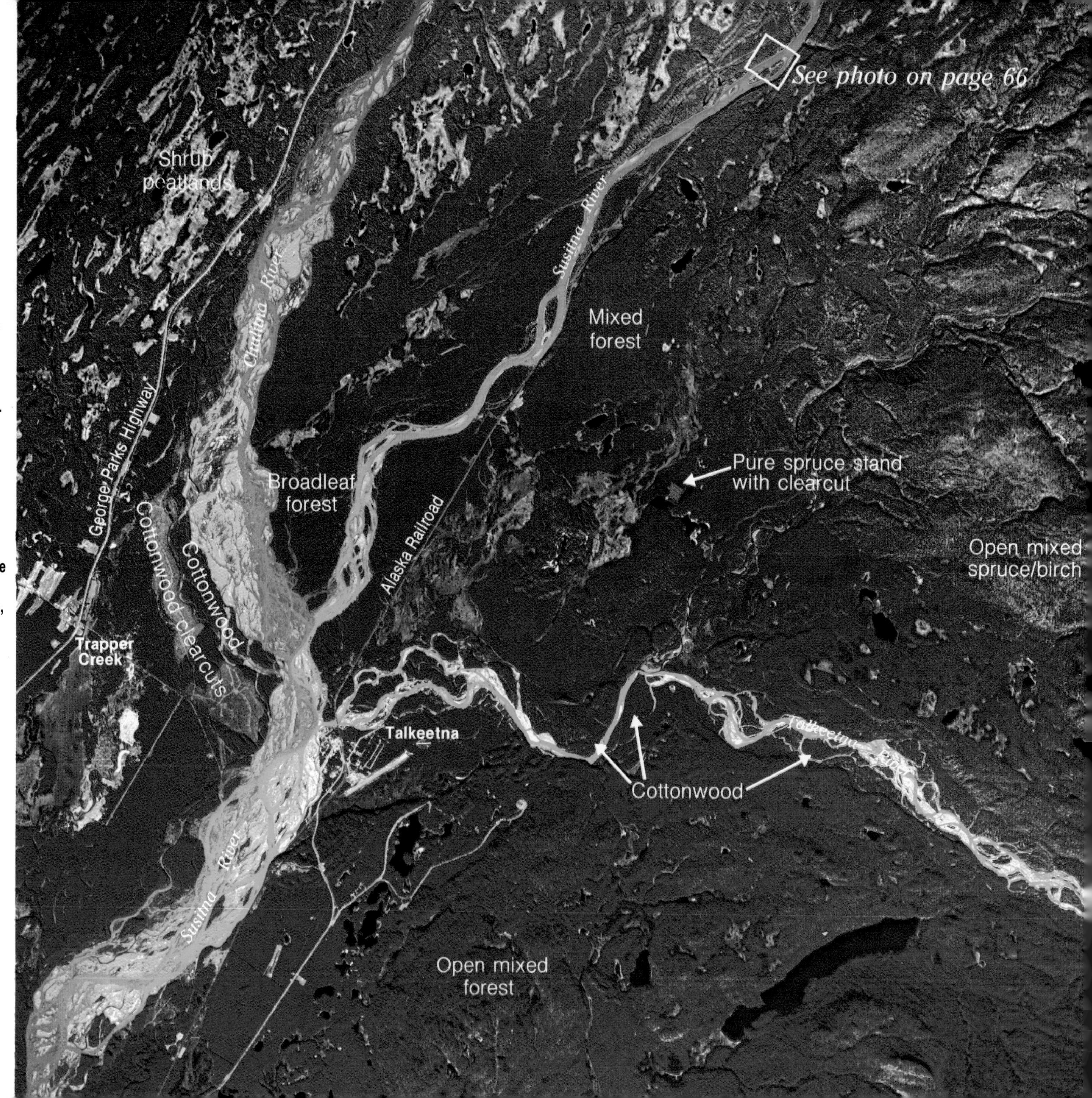

This aerial photo, taken from a NASA U-2 reconnaissance plane flying 60,000 feet over the Susitna River, is a more detailed examination of the area marked with a block on the photo at left. The George Parks Highway and Alaska Railroad are evident running vertically through the left side of the photo, and the communities of Trapper Creek and Talkeetna are visible in the left lower quadrant. This is also a false-color photo, taken with film sensitive to infrared light: bogs are dull pink-gray; shrublands are bright pink; mixed forests are muddy reds; pure spruce stands are brownish purple; and the brightest orange-red indicates healthy stands of broadleaf trees, especially cottonwoods, along the rivers.
(NASA photo, courtesy of Forestry Sciences Lab)

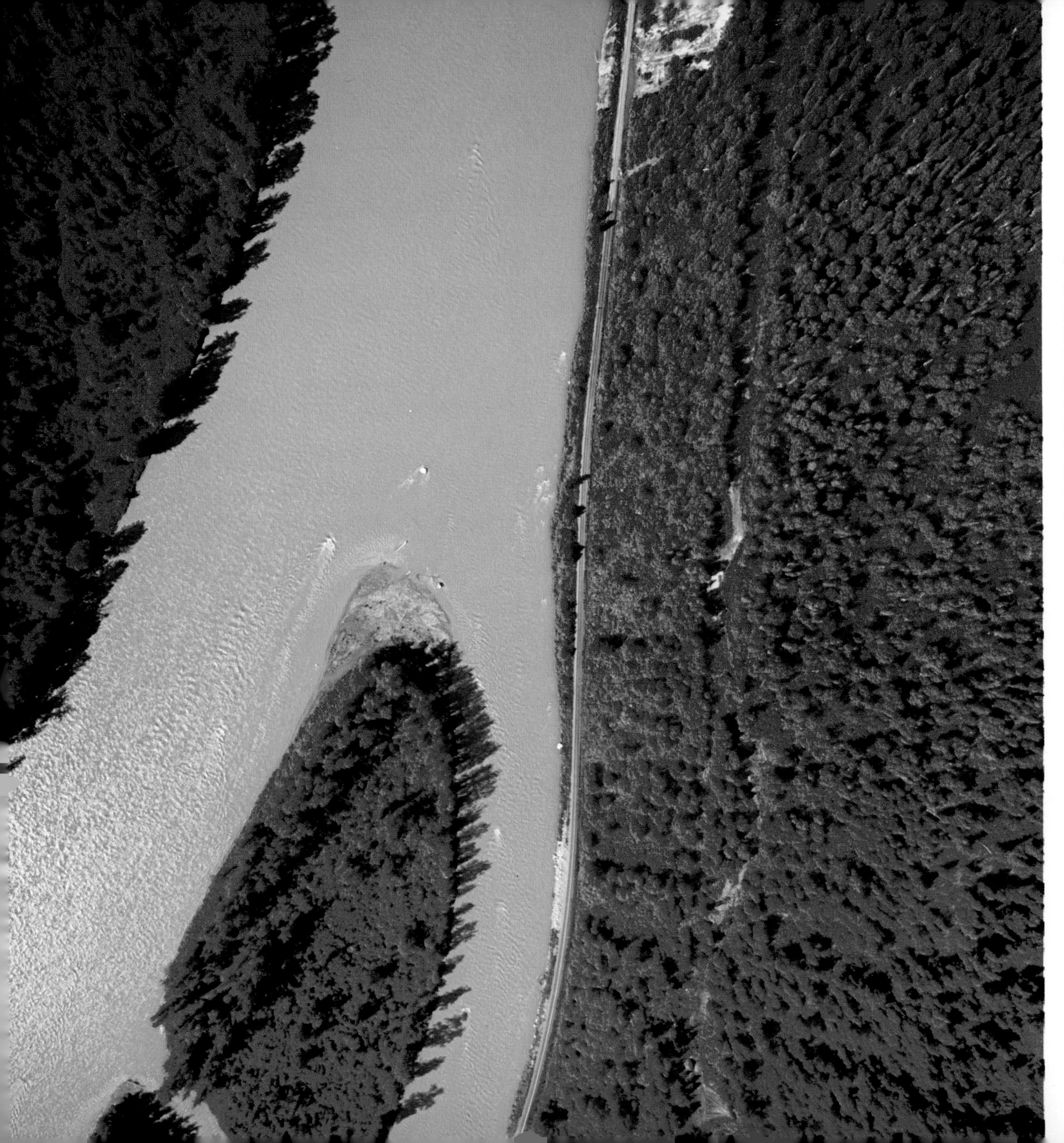

The small square marked near the top of the photo on the previous page is examined yet another time, by way of this low-altitude photo. With a scale of 1 inch = 250 feet, details such as individual trees, railroad tracks and telegraph line rights-of-way can be seen clearly. Trained interpreters can discern tree species, height and timber volume from these aerial photos.

(Courtesy of Forestry Sciences Lab)

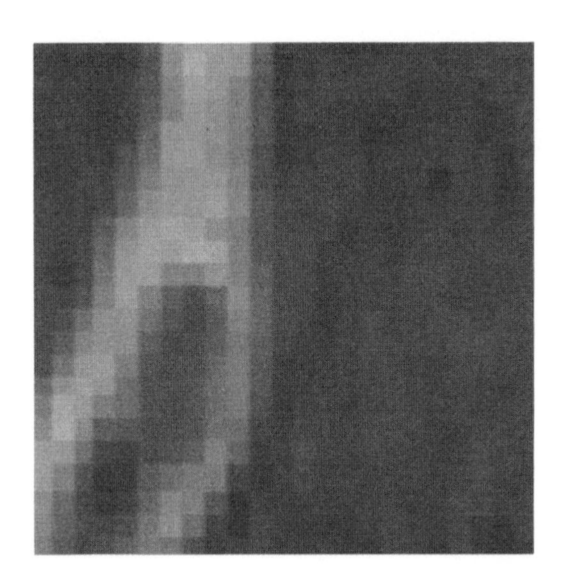
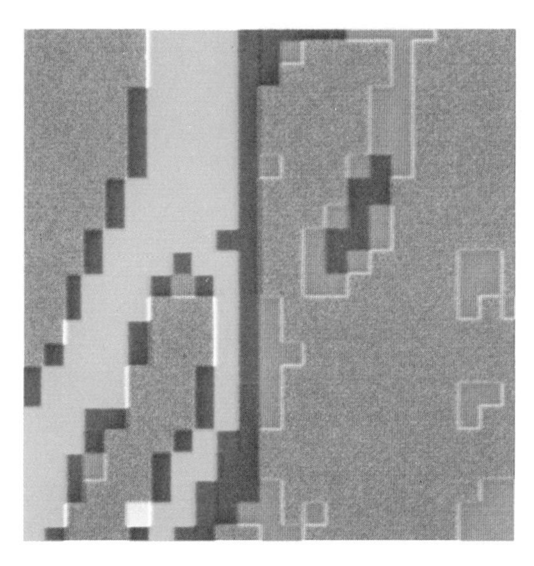

These four images are blowups, done by a computer using Landsat data, of the area shown in the photo at left. Individual pixels, each representing 160 square feet, are visible. The image at top left uses the standard false-color scheme. Top right provides information about ground cover, lumped into broad categories: tundra and alpine shrubs (yellow); broadleaf shrubs (red); broadleaf trees (orange); conifer and mixed forest (green); or water, rock, and shadow (blue). The lower left image is a more detailed view of the blue and green pixels; lower right is a closer look at the red, orange and yellow pixels. (Courtesy of Forestry Sciences Lab)

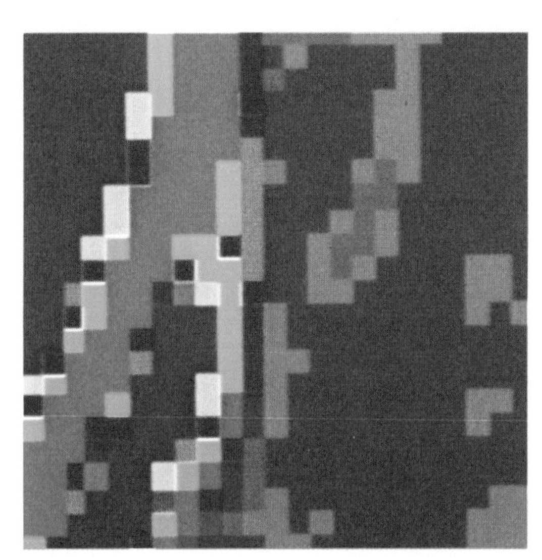
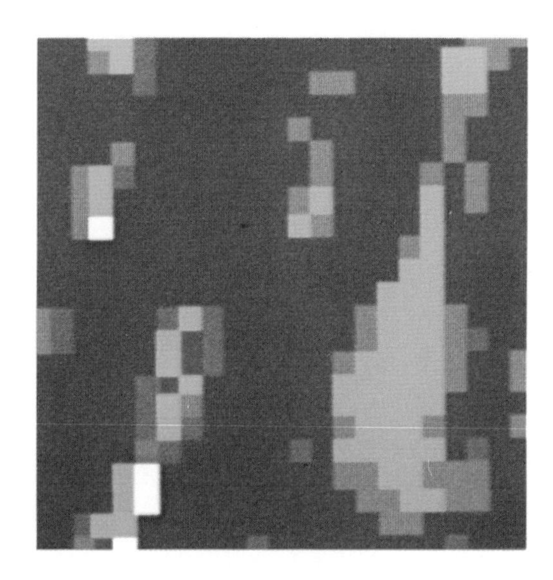

Growing Trees

Silviculture is the science of growing trees. Silviculturists believe that forests should be managed like agricultural crops for human benefit. But a forest and a field of corn differ substantially. Silviculturists recognize this, and also that the highest benefit from a stand of trees may not necessarily involve cutting it into timber.

But often, timber indeed makes the best use of a particular stand of trees. That is when the science is called upon to help grow healthier trees faster, and to best meet the anticipated needs of those who will harvest the new stand at the end of its **rotation**, perhaps more than a century from the original harvest. Silvicultural improvements require investments today that won't be paid back for a long time. Land managers may decide that they can't afford them; in effect, they may decide to let the forest grow on its own. That's pretty much how most forests have been and are being managed in Alaska. But things are starting to change.

Scandinavians and Russians have been managing their forests — many as far north as those in Alaska — very intensively for many years. The Finns, for example, have increased their growing stock volume (the total volume of all living trees of commercial species) at an annual rate of 3.9 percent. While not especially high in an absolute sense, that rate compares quite favorably with the average of only 2.6 percent for the entire United States.

An even-aged forest, containing trees of similar age and size, grows along the Glacier Highway near Juneau. (Rollo Pool)

This large, old growth spruce towers over Prince of Wales Island. (Don Cornelius)

Species grown in Alaska and Scandinavia differ, as does the climate. Scandinavia is located much closer to its timber markets than is Alaska, and landownership patterns are different from ours. Yet enough similarities remain that our foresters can learn quite a bit from their Nordic colleagues.

Three Scandinavian foresters visited Alaska in 1977 and concluded that ". . . the physical production potential for [Alaska's Interior] forests and their importance have been largely underestimated so far. Forestry, the forest industries and the services they maintain could outweigh the value of most other activities. In order to be able to utilize available possibilities, intensive research and developments are needed."

Regeneration

Some of the first forestry research in Alaska was done by the U.S. Forest Service in the 1920s and 1930s in southeastern Alaska to determine how well coastal hemlock-spruce forests regenerate after logging. Pioneer researcher Dr. Ray Taylor concluded that, in most situations, natural regeneration was adequate to reforest clear-cuts, the only practical logging method for pulpwood. Subsequent research has confirmed this finding.

In the early 1950s the Maybeso Experimental Forest was established on the east coast of Prince of Wales Island to study the effects of large-scale clear-cutting on forests and streams. In one area, called the "mile-square

Twenty-five-year-old second growth surrounds a former yarder landing near Hollis on Prince of Wales Island. (Don Cornelius)

clear-cut," a grid of small study plots and seed traps was set up immediately after logging, and was monitored on a regular schedule. Within a few years, new spruce and hemlock were growing on more than 80 percent of the Maybeso plots — a favorable percentage.

The average old growth stand in southeastern Alaska contains about 64 percent western hemlock and 28 percent Sitka spruce. But 41 percent of the new trees in the Maybeso clear-cut were spruce because the shade-tolerant hemlock lost some of its advantage over spruce, which thrives in the open. Because spruce makes more valuable **sawlogs** than hemlock, clear-cutting is also seen as a way of changing the species mix and increasing the economic value of the stand.

Twenty to thirty years after clear-cutting, a dense stand of fast-growing trees, sometimes as many as 20,000 per acre, blankets most areas in the coastal forest. Keen competition exists among trees, and the canopy closes, cutting off most light to the forest floor. Understory plants like blueberry, salmonberry, huckleberry, elder, currant, bunchberry, skunk cabbage and others that luxuriated in the open are shaded out. The forest floor becomes a dark dog-hair thicket guaranteed to raise the temper of anyone unfortunate enough to enter.

Not all areas regenerate equally well, however. On stream terrace sites and areas where the soil has been severely disturbed— such as roads and skid trails — alders become

In some areas, alders come back so well after logging that they retard development of the conifer crop. Here, silviculturist Rick Hauver applies herbicide to alders to release the spruce.
(Walt Matell, USFS)

established before conifers. The desired softwood trees can be suppressed under this broadleaf canopy for up to 80 years.

In the Interior, paper birch can produce up to 300 million seeds per acre in a good year. White spruce production varies between zero and 16 million seeds per acre. Bumper seed crops occur on a somewhat periodic basis, more frequently for hardwoods than for softwoods. Some evidence suggests that heavy seed production occurs the year following a hot, dry summer. Such years, interestingly enough, tend to be heavy fire seasons.

Quaking aspen and balsam poplar also reproduce **vegetatively**. When the aboveground portions of aspen die, suckers with extra-large leaves sprout from undamaged root systems. Birch sprouts are more clustered, since they arise from the old stumps. These wonderful adaptations help assure continued survival of the stands.

Black and white spruce can also reproduce vegetatively, by sprouting from lower branches that get buried. Conifers, however, depend mostly on seed dispersal.

Interior forests will eventually regenerate after logging, as evidenced by the **second growth** stands that today cover the hills around Fairbanks and other interior towns. These areas were harvested extensively during the mining booms at the turn of the century. The forests have grown back in natural succession, and therefore have a higher proportion of hardwoods today than the predominately spruce forests they replaced.

Reforestation surveys on upland white spruce, birch and aspen forests logged in the 1960s and 1970s indicate relatively low forest regeneration by national standards. Revegetation by grasses, ground plants and shrubs is almost never a problem. But when silviculturists examined standardized sample plots, they found as few as 8 percent and at best 73 percent containing growing tree seedlings. Since desirable restocking is thought to be at least 40 to 60 percent, research is now underway to develop methods of improving regeneration after logging the taiga.

Controlling Stand Density

In southeastern Alaska, Dr. Ray Taylor reported that "hemlock trees 18 inches in diameter are sometimes 1,000 years old and may grow not far from another tree of the same diameter which is only 200 years old."

Slow-growing trees such as these probably grew in conditions of intense competition for light and nutrients, such as deep in the second growth dog-hair thickets. Although the total volume of fiber produced in such overdense stands is high, it is distributed among many skimpy trees. For timber production, foresters prefer to have fewer trees, large and well-formed.

So, like gardeners who wish to grow prize carrots, land managers can thin such overcrowded stands. The trees that remain get

Before (above) and after photos show part of a 30-acre thinned birch stand on Toghotthele Corporation lands near Nenana. (Both by Bill Zufelt, Tanana Chiefs Conference)

Old growth stands like this one near Ketchikan are considered over-mature by foresters. (Don Cornelius)

After three years, a clear-cut begins to revegetate near Klawock, on Prince of Wales Island. (Walt Matell)

more light, and can better use the site's nutrients. After **thinning**, their growth accelerates dramatically; in silvicultural terminology, they are **released.**

Thinning is sometimes done several times in the life of a stand. **Precommercial thinning** usually takes place about the time a young stand's canopy closes; the trees removed are not yet of commercial size. Later, when the trees have grown larger, the land manager may do additional **commercial thinnings**. Managers prefer to have the value of the trees removed cover the cost of the thinning operation.

The Forest Service has been precommercially thinning about 6,000 acres annually in Tongass National Forest. In interior Alaska, thinnings on research plots have shown favorable results with white spruce and aspen. Toghotthele Corporation has thinned 30 acres of birch on its lands near Nenana.

Silvicultural Systems

Until recently, harvesting timber was commonly seen as the final step in managing forests for wood and fiber. Silviculturists like to point out that cutting, which can be done in several different ways, is also the first step in starting a new stand. The appropriate silvicultural system depends on the site, the species and economics.

The most common system of cutting timber in Alaska today is **clear-cutting**. This involves cutting down all the trees in a given

area at one time. Large areas, up to several square miles, have been harvested this way in the past, but today most clear-cuts on public land cover less than 100 acres (less than one-sixth square mile). A **drainage cut** is a clear-cut that harvests most of the commercial timber in a valley at one time. **Patch-cuts** are a series of smaller clear-cuts separated by uncut forest.

Clear-cutting opens up the forest floor to light, making it easier for species that need light, such as Sitka spruce, to regenerate. It also warms forest soils, speeding up decomposition of debris into nutrients. In parts of the Interior subject to hot, dry spells during the summer, unprotected seedlings can be killed. Conversely, the excellent regeneration in southeastern Alaska clear-cuts is aided by the more moderate summers there. The increased light also stimulates dense growth of shrubs and cover plants, which remain in the clear-cut until the second growth canopy closes.

Clear-cutting is also the most economical harvest system, as it allows quicker access to more forest and requires fewer roads than other methods. Generally, larger clear-cuts yield greater immediate economic returns.

Silviculturists study forest regeneration by observing sample plots such as this one in the Bonanza Creek Experimental Forest west of Fairbanks. Above left is a one-square-meter area after shelterwood was cut in 1973; at left is the same area five years later. White spruce seedlings are hidden under birch, the first tree species to grow back in this area. (Both by John Zasada, Institute of Northern Forestry)

The **seed-tree system** is essentially like clear-cutting, except that some trees are left standing to disperse seeds over the cut. This method is appropriate in large clear-cuts or areas with low natural regeneration, and can be used to select the species the landowner favors as second growth. The seed-tree system does not work successfully, however, in areas with stormy weather that would blow down the isolated seed trees.

With the **shelterwood** system, the forest canopy is partially opened up in successive cuttings, leaving enough large trees to provide both seeds and protection for the regenerating stand. Once the second growth has become established, harvesting the remaining shelterwood releases the new trees.

Shelterwood is not used in the coastal forest because it costs more than clear-cutting, and usually doesn't offer any silvicultural advantages there. The system has been tried on the Bonanza Creek Experimental Forest west of Fairbanks, with regeneration not much different from that on adjacent comparison clear-cuts.

The **selection system** of harvesting, unlike the others, is designed to maintain an **uneven-aged** stand in perpetuity. As often as economical, individual or small groups of trees are carefully removed from the stand.

◄**Several tree species form flowers on catkins, long clusters named for their resemblance to a cat's tail. The catkin at left hangs from a balsam poplar; at lower left, an aspen catkin releases its cottony seeds.** (Both by John Zasada, Institute of Northern Forestry)

▼**Perfectly formed cones are clustered on this white spruce bough.** (Bob & Janet Klein)

Interior foresters are fond of talking about "Herm's Tree," a white spruce cut down in 1977 and named after researcher Francis R. Herman. By counting growth rings in cross sections of the 22-inch diameter trunk, foresters discovered that the tree was 208 years old. The rings were tightly spaced until 1917, when the tree was five inches in diameter and about 150 years old. Evidently, surrounding trees were logged about that time, because in the remaining 60 years, it added 17 inches of new growth. Here, forester Dean Argyle points to the spot where the growth rings suddenly become much wider. (Walt Matell)

The primary aim is to improve the stand, so diseased or defective trees are removed first in unhealthy stands; later, improved-quality trees are harvested on a regular basis. (Selection cutting should not be confused with **selective cutting**, which involves "high-grading" the best trees and leaving the trashy ones.)

Selection cutting requires many roads and is expensive. It is used in areas like campgrounds, scenic areas and special wildlife habitats, where managers desire to maintain the characteristics of uneven-aged timber. In forests of thin-barked species, like spruce and hemlock, frequent logging required by the system increases the chance of damage to the remaining trees.

Helping the New Stand

Logged areas in Alaska will often regenerate naturally from seed and seedlings left on the ground, or from seeds blown in from adjacent stands. But in some areas — particularly in the Interior — although seeds are available, few seedlings survive due to poor growing conditions. Foresters are therefore placing much emphasis on **site preparation**: creating optimum growing conditions to get the new crop of trees established.

Tree seedlings have difficulty growing under dense layers of brushy vegetation that quickly invade cutover areas. They also don't grow well in the cold, thick organic mats

A tractor pulls a Scandinavian scarifier through a recently logged experimental plot at Willow Island near Fairbanks. The machine prepares the site for regrowth by gouging surface organic material down into the mineral soil. (Walt Matell)

Forest geneticist John Alden holds a containerized white spruce seedling about to be planted in an experimental regeneration plot west of Fairbanks. The burned plot will be planted with seedlings grown from seed gathered at various altitudes to determine the best seed source for this lowland site. (Walt Matell)

covering the forest floor in much of the taiga. Site preparation, therefore, includes brush control and breaking up the organic layer, which is done mechanically with specialized equipment, or by using fire.

Silviculturist John Zasada with the Institute of Northern Forestry believes that **prescribed fire** is the ultimate tool for site preparation. It best simulates the effects of natural fires, with which the northern forest has evolved. He cautions that such burning should be done very carefully to avoid starting non-prescribed wildfires.

Planting and Fertilizing Trees

When nearby seed trees are not available, land managers can consider seeding or planting seedlings. Seedlings are grown under controlled conditions in nurseries in the Pacific Northwest, and in two new Alaskan facilities: the Forest Service's B. Frank Heintzleman Nursery in Petersburg and the Alaska State Forest Nursery in Eagle River. Both grow a variety of species for reforestation and research.

Planting can also be used to give desired species a head start. Sitka spruce have been replanted on some areas cut by Sealaska Timber Corporation in southeastern Alaska. The forest management firm, Koncor, planted some 1.1 million Sitka spruce seedlings on Afognak Island. And Toghotthele Corporation has planted 21,000 white spruce on its taiga lands near Nenana.

If we can chose what trees to replant, then how about planting imported species that have desirable growth and timber characteristics? The British and Icelanders have done this, and today have forests of Sitka spruce imported from Alaska. The Scandinavians have planted Canadian lodgepole pine, and anticipate higher timber production on a shorter rotation.

Forest researchers urge caution before investing money — and land — in exotic forests. A few imported lodgepole pine seedlings have been planted on an upland test plot at Bonanza Creek Experimental Forest, and are doing quite well after 10 years. But it is too early to tell whether lodgepole pine will survive to maturity where we know white spruce and paper birch thrive.

Other experimental introductions into Alaska include: Norway spruce, Siberian larch, Scotch pine, Siberian fir and subalpine fir in the Interior, and Douglas fir in southeastern Alaska.

Besides light and moisture, trees need the right kinds of nutrients for proper growth. Silviculturists interested in stimulating tree growth consider supplemental fertilization, especially since many forest soils are deficient in available nitrogen, a key nutrient.

In the Interior, a 15-year-old aspen stand that was test-treated with varying fertilizer combinations yielded promising results: new growth on some of the fertilized trees reached 12 times as much as on unfertilized control

Joe Stehlik, manager of the State of Alaska's Eagle River nursery, checks the progress of some of the 600,000 seedlings the facility grows each year. Most are native conifers, but the nursery also raises some hardwoods and exotic species. This bed contains mostly lodgepole pine; the smaller trees in front of Joe are white spruce; and the pale cluster in the center is a flat of tamarack, the only Alaska conifer that loses its needles each year.
(Walt Matell)

State Division of Forestry personnel measure bushels of white spruce cones collected near Fairbanks. Cones are identified by area gathered and are then sent to the Eagle River nursery, where they are seasoned in a controlled environment.
(Walt Matell)

Russian 1805 Plantation on Unalaska

After the naturalist Adelbert von Chamisso visited Unalaska Island with the von Kotzebue expedition in 1816, he reported that "Oonalashka, and the other islands in the chain, are entirely destitute of [trees]. It has been attempted to plant pines, a kind of *Abies*, brought from Sitka, at Oonalashka; most of them have perished, the others seem scarcely to thrive, but the plantation is still young, and it is well known how ill coniferous trees bear transplanting."

Chamisso's description was the first known reference to a plantation of Sitka spruce (*Picea sitchensis*, not a pine or fir as Chamisso described) still growing today near Dutch Harbor on Unalaska Island. In 1834, Ivan Veniaminov reported that 24 trees were still alive, and that some were more than seven feet tall. Planted in 1805 by the Russians, it may well have been

Evidence remains today of the limited success of the 1805 Russian experiment to transplant Sitka spruce onto windswept Unalaska Island. Progress of the plantation, located near Dutch Harbor, has been documented through the years by photographs; the trees are shown in the photo at far left in 1899 and at left in 1978.
(Far left, National Archives, courtesy of Richard Tindall; left, Richard Tindall)

the first attempt at afforestation in America.

The windswept nature of the island — located well beyond the natural tree line — is evident by the low, bushy form of the plantation's trees. Forester Richard W. Tindall, who has visited Unalaska Island frequently to study the plantation, reports six trees surviving. The trees average 37 feet tall (the tallest is 43 feet) and have an average trunk diameter of 21.4 inches at breast height.

The Russians planted a few other plantations on Unalaska. More than a century later, during World War II, the U.S. Army followed this tradition, planting seedlings throughout the Aleutian chain with varying degrees of success.

In Petersburg, seedlings are grown at the Forest Service's B. Frank Heintzleman nursery for reforestation and research. (USFS)

trees, and diameter growth increased up to 4 times.

The Forest Service has fertilized test plots in southeastern Alaska and on Afognak Island. In most of these experiments, the fertilized trees grew faster than trees on similar control plots.

Although such results look promising, fertilization can have its drawbacks. While applying the correct amount to the trees, silviculturists must take care to avoid dumping any fertilizer in lakes and streams. And resulting tree growth isn't free; with regular fertilization, extra growth must more than pay for the added expense.

Rather than applying man-made fertilizer, silviculturists may use plants that fix nitrogen into the soil, like alder. Researchers in the Pacific Northwest think that a short rotation of red alder between Douglas fir timber crops compares favorably with artificial fertilization. Such research is needed for Alaskan species. Another possibility for future forests is to innoculate them with **mycorrhizae**, fungi that help the trees' roots absorb nutrients.

Protecting the Forests

When we manage a forest for timber or for other human uses, we intervene in the constant struggle between the tenacity of trees and the forces that seek to destroy them.

On Thanksgiving Day, 1968, a storm with hurricane-force winds gusting to 90 miles per hour took only six hours to pass through Alaska's Panhandle. One billion board feet of timber lay in its wake, blown down as if in a gargantuan game of pick-up-sticks.

Winds have a tendency to stream over the tops of a closed forest canopy and dip down into openings, so clear-cuts in windy areas should be planned with care. Units can be located to take advantage of naturally windfirm boundaries along muskegs and other openings, where the trees gradually increase in height, creating a wedge facing the wind. Foresters can also design clear-cut boundaries to minimize exposure of vulnerable edges to prevailing high winds.

Weather can also damage forests in the Interior. In 1968, a storm broke about one-quarter of the snow- and ice-laden spruce in a stand in the Bonanza Creek Experimental Forest.

While winds come and go overhead, insects and fungi munch on the forest quietly and constantly. Whether they kill trees outright, or produce defects that lower the timber value, these agents greatly concern the landowner managing a forest for wood or fiber.

The spruce bark beetle (also known as the spruce beetle) tops most lists of pests. In the 1940s, this insect killed 35.5 million board feet of prime Sitka spruce on Kosciusko Island in southeastern Alaska.

In 1982, the beetle infested almost one-half-a-million acres of forest in southcentral Alaska and at Glacier Bay. The largest outbreak was

Les Viereck of Fairbanks stands next to the sign which designates his property the most northerly certified tree farm in the United States. Any landowner who actively manages at least 10 acres of forestland is eligible to apply for membership in the American Forest Institute's Tree Farm System. Viereck has done some thinning on the 20-acre farm, and plans to maintain a greenbelt for aesthetics and wildlife habitat. (Tony Gasbarro)

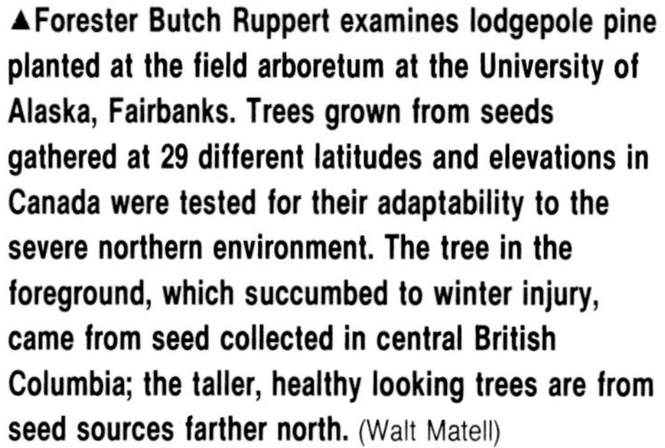

▲Forester Butch Ruppert examines lodgepole pine planted at the field arboretum at the University of Alaska, Fairbanks. Trees grown from seeds gathered at 29 different latitudes and elevations in Canada were tested for their adaptability to the severe northern environment. The tree in the foreground, which succumbed to winter injury, came from seed collected in central British Columbia; the taller, healthy looking trees are from seed sources farther north. (Walt Matell)

►This blowdown, along the edge of a clear-cut, near Nakuti on the west coast of Prince of Wales Island occurred during a storm. One cause of blowdown is thought to be shallow rooting that results because of a hard, impermeable iron pan which forms in the top layers of mineral soil. Paradoxically, the falling trees' roots churn the soil, breaking apart the iron pan, creating a more favorable seedbed for new trees. (Tom Kogut)

near Beluga Lake on the northwest shore of Cook Inlet.

The beetles kill spruce by burrowing beneath the bark, destroying the phloem, the thin layer of living tissue that transports food from the needles to the tree's roots. The beetles congregate by sensing chemical attractants, **pheromones**, that they release. Small populations of beetles, usually present in dead or slow-growing trees of a healthy spruce forest, are kept in check by weather, woodpeckers and parasites. When conditions favor the beetle, their numbers increase suddenly and they may attack living trees as well.

The most effective control is to dispose of infected trees and downed material, and to avoid accumulations of slash that might get infested. At Tyonek, south of Beluga Lake, a state timber sale was offered in 1973 to salvage white spruce killed by the spruce beetle. This operation closed in 1983, after the operator harvested 74 million board feet of damaged spruce.

These Alaska-cedars near Slocum Arm on Chichagof Island are dying. About 25,000 acres in southeastern Alaska are affected, prompting extensive research by scientists to discover the cause. Plant pathologists have virtually ruled out insects; they are now trying to isolate and identify fungi, and are also considering other factors such as environmental stress. (Terry Shaw, Forestry Sciences Lab)

The large aspen tortrix caterpillar feeds on aspen leaves, and when present in epidemic numbers, defoliates the tree. It infested 26,000 acres near Willow in the late 1970s, and a similar-sized area at Point MacKenzie in 1982.

Aspen has the capacity to grow new leaves shortly after defoliation, so one attrack rarely damages the stand. Even after three years of heavy defoliation at Willow, only 16 percent of the trees died, although such attacks did weaken the stand. Tortrix populations usually decline naturally, and do not normally require control measures. Taking care not to damage the roots or stems of infested trees, and perhaps fertilizing them, bolsters the aspen's inherent vigor.

In the 1950s, almost one-third of the net volume of western hemlock on some sites in southeastern Alaska was lost to the western black-headed budworm. After the eggs overwinter on their host, the budworms emerge in the spring to feed on new growth. Later, if present in epidemic numbers, they will defoliate the entire tree. Budworm populations seem to depend on weather, and decline as rapidly as they increase.

The tiny green spruce aphid defoliates all species of spruce, but prefers Sitka spruce. Abnormally high populations infested 21,000 acres of southeastern Alaska in 1981, particularly near Ketchikan. Kodiak Island also had an outbreak in 1981, but a cold snap the following winter checked the epidemic. Although some trees near Sitka have died

In terms of damage, the spruce bark beetle (above) is probably the major pest in Alaska's forests. The beetle kills spruce by burrowing under the bark and destroying the layer of tissue which carries nutrients from the tree's needles to its roots. The beetle-infested spruce below has had its bark removed by woodpeckers.
(Both by Ed Holsten, USFS/S&PF)

This western hemlock stand on Dall Island shows damage done by hemlock sawfly infestation.
(Don Cornelius)

after aphid defoliation, most trees infested with the insect survive.

Ambrosia beetles thrive in dead trees and downed hemlock, Sitka spruce, birch and white spruce, boring elaborate egg galleries into the outer sapwood. They often infest wind-thrown stands, such as the 100 million board feet that blew down near Yakutat in 1981. Similar infestations have been found in newly logged timber, log rafts and even on spruce and hemlock **cants** at mills in southeastern Alaska. Many buyers will not even consider logs damaged by this insect.

An interesting control measure for ambrosia beetles is to lure them into traps baited with synthetic pheromones.

Also dining on Alaska's forests are other insects with colorful names like the birch leaf roller, blotch miner, hemlock sawfly, four-eyed spruce bark beetle, gall midge and various loopers, among others.

Hemlock dwarf mistletoe, found throughout the northern end of western hemlock's geographic range, has been termed "one of the major forest pests in southeastern Alaska." This parasite is actually a seed-bearing plant which roots on the branches and stems of hemlock. When it does, the host limbs swell and the tree's vigor is reduced. Four to seven years after attachment, the plant flowers and pollinates. When the fruits mature they explode, ejecting sticky seeds 30 to 50 feet onto adjacent trees. While mistletoe rarely kills its host, it can reduce tree growth

The two seedlings on the left show successful innoculation with a mycorrhizal fungus; the seedlings on the right are untreated. The beneficial fungi help the tree's roots absorb nutrients. (Terry Shaw, Forestry Sciences Lab)

and create entry points for decay-causing organisms.

Severe mistletoe infestations are found mostly in old growth stands. Fortunately, mistletoe appears to spread slowly in younger second growth; stands managed on an 80- to 100-year rotation will probably suffer little damage. Neither chemical nor biological controls have yet been developed. The best strategy for controlling mistletoe seems to be **sanitary cutting**: harvesting infected stands and removing or destroying infected trees.

When we see fungus **conks** — the fruiting bodies of decay fungi — on trees, we are alerted that something rotten is going on inside. Sometimes, as in the case of tinder fungus, which infests birch, the conks do not appear until little heartwood remains in the tree. Precautions against decay include protecting trees from physical damage, since the fungi must first pass through the trees' protective bark.

Forest disease surveys have shown that while brown and white rots quite busily recycle old growth trees, they rarely occur in young, second growth stands. Plant pathologists are concerned that the anticipated increase in thinning of second growth may leave thousands of exposed stumps and tree scars. Research is now under way to determine whether the second growth stands might become infected through these openings.

As trees mature, they become more susceptible to disease. In southeastern Alaska, about 32 percent of the gross volume of old growth stands cannot be used as sawtimber, with Sitka spruce having the least defect (9 percent)

Conks such as this "chicken of the woods" *(Polyporus sulphureus)*, growing here on western hemlock, are the fruiting bodies of fungi that decompose the inside of the tree. (Walt Matell)

A large burl grows on a white spruce along the Susitna River. Although silviculturists do not know what causes burls to form, they do have some theories, such as growth over an irritant, similar to the formation of pearls in oysters. The defect that burls cause in wood lowers the value for lumber, but increases its value for decorative items such as coffee tables. (Don Cornelius)

Bird pecks around bear scratches on this alder indicate it has had rough times, and will probably die from fungal infection that can enter through these openings. (Ron Bonar, USFS)

and cedar the most (52 percent). When foresters refer to old growth stands as decadent or overmature, and younger stands as vigorous and thrifty, they are talking about this increased defect.

Animals not only live in the forest, they also eat it. Mice, voles, shrews, squirrels and other rodents consume seeds or seedlings. During severe winters, deer and moose will browse seedlings and even **saplings**. Porcupines kill trees by eating the inner bark and girdling the tree in the process. Beavers can thin a stand to excess. In a healthy forest, these creatures are kept in check by fluctuating food supplies and predators. If the

Porcupines feed mostly on the inner bark of spruce, birch and aspen. At times, they will spend days or weeks in one tree, methodically gnawing at the bark and usually killing the tree. (Lisa Holzapfel)

system breaks down and some populations get too large, the forest could be damaged.

Some interesting relationships have evolved between animals and trees. Birch, for example, develop repellent resin glands on lower branches which are threatened by hares; upper branches out of the browsers' reach don't develop this protection. Mice and voles eat underground truffles, and thus spread spores of beneficial mycorrhizal fungi that would not otherwise be distributed.

Growing Tomorrow's Forests

Silviculture in Alaska is relatively young, having been practiced for less than an average rotation. There are many problems waiting to be solved. Landowners interested in managing forests for wood, fiber and energy want to know, for example, why hemlock flutes, and if it is possible to grow hemlock without this defect. Can production be improved on muskegs and boggy areas in Alaska by draining them, as is practiced in Finland? If so, what are the ecological consequences of this? How are nutrients and energy recycled in the forest, and how can they be harnessed?

These and many other questions will take time to answer, as silviculture is long-term agriculture. And even when we have the technical answers, we will still have to determine whether treatments are economical, and how they mesh with other resource considerations.

Traditional Uses of Alaska's Forests

Alaska's original inhabitants were as diverse as the lands they populated, but they had one thing in common: whether they lived in temperate coastal forests, on the barren Arctic coast or along wide interior rivers, they developed methods of survival — cultures — that relied on ingenious use of available resources, including trees.

Trees rarely grew along the northern and western coast of Alaska, so Eskimos living there valued the driftwood delivered to them by rivers and ocean currents. They also made forays up river valleys to secure wood, sometimes from shrub species like willow.

Wood supplemented whale bone in frames for boats such as the large open *umiaks*. Eskimos carved wooden bowls, tools, masks, figurines, grave markers and other objects. Driftwood was used to construct houses; frames for skin tents were made from willow. Eskimos ate young birch leaves and the inner bark of willow; they made a chewing gum from spruce sap. Willow roots became sinews to tie things together, or were woven into baskets. Willow bark was twisted into rope and fishing nets. Baskets, containers and canoes were made from birchbark.

Beyond tree line in their island realm, the Aleuts built houses called *barabaras* half-underground for warmth. Driftwood poles set in the corner supported roofs covered with earth and sod. Occasionally measuring up to 40 by 240 feet, these dwellings accom-

An Eskimo of the lower Yukon stands next to an elaborate wooden fish trap; his catch dries in the background on a wooden rack.
(C. L. Andrews Collection, Alaska Historical Library)

►**An Ahtna Indian family from Taral, on the east bank of the Copper River, poses in the late 1800s. Their camp includes a shelter and fish-drying rack constructed from local trees.** (Alaska Purchase Centennial Collection, Alaska Historical Library)

▼**Ind-a-yanek, the "best known native guide in Alaska" in 1907, stands proudly with a pair of expertly crafted snowshoes.** (Laura M. Hills Collection, University of Alaska, Fairbanks)

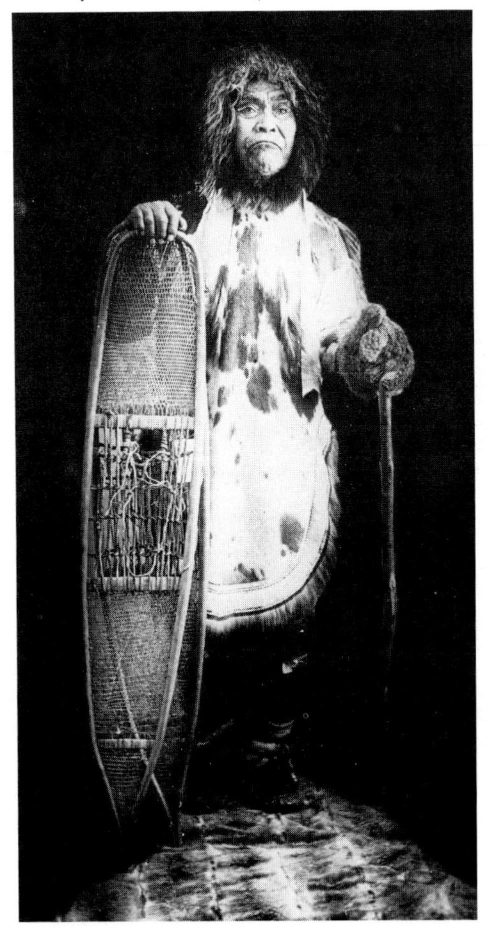

modated several families at one time. Aleuts also made frameworks for their elegant sealskin boats — *bidarkas* — from carefully selected driftwood.

Along interior rivers, inland Eskimos and Indians made elaborate fish traps from willow and split spruce, setting them at the ends of stick fences for best success. Sleds and snowshoes, necessary for winter travel, were made of various woods; summer travel was often by birchbark canoe. Semipermanent houses were constructed of caribou or moose skins stretched over dome-shaped wooden frames, or of logs similar to the white man's log cabin but without fitted cuts. Athabascans sometimes made summer houses with spruce or birchbark walls.

Wood and salmon formed the material basis of the culture of southern and southeastern Alaskan peoples, considered by anthropologists to have been the wealthiest of Alaska, if not of all North America.

Native women paddle their birchbark canoe along the Yukon River. (Charles Bunnell Collection, University of Alaska, Fairbanks)

Early white explorers reported houses built of logs big and long enough to have made mainmasts for ships of the British Navy. Planks, split by maul and wedge, were secured as walls and roofs to stout posts and beams, forming large communal houses. House posts were often elaborately carved, as were freestanding totem poles.

Master boat-builders constructed dugout canoes up to 70 feet long from cedar logs. After carefully burning and adzing the logs, they widened the sides of the craft by softening the insides with heated salt water. Fishing hooks, paddles, harpoons, storage boxes and other wooden implements were elegantly designed, and often decorated. Women wove spruce roots into tight baskets capable of holding liquids. Cedar bark provided fibers for clothing and, combined with mountain goat wool, formed the warp of the famous Chilkat blankets. Masks, rattles, hats, drums and other ceremonial objects were also made with materials gathered in the region's abundant forests.

The forests of Alaska remained abundant because use was extremely light and dispersed. The natural forces of a harsh environment kept human populations in check. The white visitors soon to arrive brought with them concentrated energy sources enabling people to live in greater numbers even in Alaska. Along with new technologies, they also imported different attitudes toward the forests and the land.

◀The Haida Indian village of Old Kasaan is shown deserted in the early 1900s. Kasaan is Tlingit for "pretty town," a name bestowed on the village because of its many ornately carved totems. Old Kasaan was abandoned in the 1890s when the settlement moved eight miles north. Several of the totems were later relocated to Ketchikan, where they have been preserved.
(C. L. Andrews Collection, Alaska Historical Library)

▼Don Bell, manager of the Alaska Loggers Association in Ketchikan, carves walking sticks from diamond willow. The unique patterns occur in several species of willow, notably Bebb willow, as a result of a fungus which grows at stem junctions.
(Kirsten Held)

Tlingit carver Nathan Jackson of Ketchikan carves a totem using traditional tools. (Robert Hallinen)

◄**An Eskimo from Ambler harvests birch bark near Fairbanks. The bark is made into buckets, baskets and canoes.** (Dean Argyle, USFS)

Carrol Phillips taps a paper birch for the sap he uses to make birch syrup. (Dean Argyle, USFS)

Ernestine Glessing, an artist from Hoonah, displays a spruce root basket she is weaving. The craft demands patience and knowledge of how to collect materials from the forest. (Walt Matell)

Russian-American Forests

by Gary Candelaria

[**Editor's note:** *Gary Candelaria is chief ranger at Sitka National Historical Park.*]

Vitus Bering's discovery of Alaska in 1741 began Imperial Russia's 126-year involvement with the people, geography and natural resources of the North Pacific.

The Russians' primary interest was fur, but the new land's timber interested the colonists from the start. The first permanent Russian settlement in North America was founded at Three Saints Bay, on Kodiak Island, in 1784. Soon after Alexander Baranov arrived in Alaska, he moved the outpost to Pavlovsk Harbor, a place chosen, in part, "since the place was surrounded by wood."

Wood was not always free for the taking, as the Russians soon discovered. In 1799, Baranov established Redoubt Saint Michael, a wooden citadel hacked out of the forest on Sitka Sound. Three years later, Tlingit warriors stormed out of the surrounding forest, destroyed the outpost, and drove the Russians away. Baranov returned in 1804 with reinforcements, confronted the Tlingits in their wooden stockade, and forced them to evacuate. This was the start of New Archangel, soon to be the capital of Russian America and later to become the city of Sitka.

The Russian-American Company, which held the trading monopoly in the colony, needed ships for its operations. The first ship built on the western coast of North America was the *Phoenix*, built at what we today call Resurrection Bay. Launching this three-masted ship reportedly ". . . gave Baranov real

This 1930s photo shows second-growth forest regenerating an area cut by Russians for charcoal nearly one hundred years earlier.
(Raymond F. Taylor, USFS)

pleasure to see timber from this far-off American wilderness being used for the good of the fatherland."

Shipyards were eventually established at Pavlovsk Harbor, Slavorossiya (Yakutat), New Archangel and Fort Ross in California. New Archangel became the shipbuilding center for the colony: four ships were built there by 1810, and ten more by 1834.

The ships had hemlock bodies, Alaska-cedar ribs and spruce decks. Spruce and hemlock were not durable ship woods, so the Russians eventually discontinued building ships, instead purchasing pine and oak vessels from Yankee traders. Occasional smaller boats were still built from local woods, and spruce and hemlock sap were made into a waterproof gum used in place of tar and pitch. The Russians also furnished wood to their Aleut hunters for the construction of paddle-propelled *bidarkas*.

More than a thousand trees were felled to build the stockade around New Archangel. Work parties of 15 to 20 soldiers were assigned to logging details, working and living in the woods for a month or two. Despite the dangers and hard work, Captain Golovin of the Imperial Navy reported that it was a popular assignment among the men.

In later years, Tlingits were hired to supply firewood for New Archangel's hungry stoves. The settlement burned 2,300 cubic *sazhens* (789,000 cubic feet) of firewood in 1861. Wood was also used to produce charcoal for

The old Russian sawmill at Sitka, shown here in 1885, operated intermittently from around 1840 through the end of the century. The mill was run by water taken from Swan Lake and carried by the flume on the left and dropped onto the water wheel. (Alaska Historical Library, reprinted from *ALASKA GEOGRAPHIC®*)

heating and foundry use. The remains of charcoal pits can still be seen in the hills behind Sitka.

The Russians constructed sawmills at Kodiak, Kenai, New Archangel and Redoubt Lake (near Sitka). Operating constantly, the mills barely kept up with the demand for lumber, and little was available for export. Records indicate, however, that a limited amount of timber was shipped to California, Mexico, Chile, Australia, the Sandwich Islands (Hawaii) and even China.

The mills also produced sawdust, which was used to insulate ice shipped from Alaska to thirsty San Francisco at the height of the California gold rush. As much as 200,000 pounds of ice was shipped annually, from both Kodiak Island and New Archangel.

But the logistics and expense of logging, and stiff competition from other suppliers — especially the Pacific Northwest — nipped any beginnings of a timber industry in the colony. If they had been successful, Alaskan history might well have been different, because with the decline of the fur trade, the Russians were anxious to enter other commodity markets to keep the Russian-American Company profitable. As it turned out, nothing was able to change the company's dismal financial outlook. The Imperial government became convinced that their American colony was the source of more trouble and expense than it was worth. So the stage was set for liquidation of the company's holdings and the sale of Alaska to the United States in 1867.

Remains of 19th century charcoal mounds are still evident in Sitka, where Russians used wood to produce charcoal for heating and foundry use.
(Gary Candelaria)

Attesting to the vastness of the land, the Russians had barely dented the forest's bounty. After these first white settlers departed, a mostly pristine land awaited arrival of the next group of users, the Americans.

Pick, Hook and Ax

William Dall, an early explorer of Alaska, was evidently happy about its purchase by the United States, and optimistic about future development of the territory. Writing in 1870, he described the forested coast: "The resources of the southern Sitkan District lie apparently entirely in its timber. This is unquestionably needed on the Pacific Coast, and a most valuable acquisition. No better lumbering district can be imagined . . ."

For the remainder of the 19th century, however, lumbering in Alaska was thwarted by government policy. Except for those on patented mining claims, all trees belonged to the federal government, which invoked a rule prohibiting export of any timber from the territory. Eighteen years after Dall's comments, Governor Alfred P. Swineford reported six sawmills in Alaska, and that "under existing conditions very few persons care to invest their means in the erection of lumber mills, the regulations [are] too ironclad, and as a consequence by much the larger half of the lumber consumed in the territory is imported from Oregon and Washington Territory."

Timber harvesting was prohibited on public domain lands — in effect, most of Alaska. Many locals cut anyway, and then settled "innocent trespass" fees with sympathetic government agents. In many areas, however, there were no agents.

Much logging during that time supported

Timbers are piled high at the Eska coal mine, in the Matanuska Valley, in 1918. Wood was used to fuel steam locomotives and in the steam-driven engines of the mine's power plant.
(Alaska Engineering Commission Collection, Alaska Historical Library)

The sternwheeler *Columbian* "woods up" along an interior river. Riverboats traveling upstream required about two cords per hour. (Charles Bunnell Collection, University of Alaska, Fairbanks)

Cords of wood line the bank of the lower Yukon River at Russian Mission in about 1900. Villagers prepared the wood for sale to passing steamboats, whose voracious boilers could consume 30 to 50 cords per day. (R. K. Woods Collection, University of Alaska, Fairbanks; reprinted from *The ALASKA JOURNAL®)*

the mining and fishing industries. In the Interior, miners cut local forests for house logs, firewood and mining timbers. Digging through permafrost was next to impossible, so miners started huge fires to thaw out their excavations. Alternatively, pipes were driven into the ground and water heated by wood-fired boilers was pumped through the pipes to thaw the ore-bearing ground. Miners were notorious for wasteful cutting, and for starting forest fires.

Along the coast, canneries and salteries used local timber when possible. Hemlock was a favorite for pilings, cedar was used to build boats, and spruce was made into fish boxes and barrels. Exporting fish in containers made from wood from public domain lands was technically illegal; in 1886, for example, the federal government prohibited the cannery at Klawock from shipping its products in boxes made at their sawmill, forcing them to import box lumber from the Pacific Northwest.

Timber companies in Washington and Oregon profited greatly from such trade. They exported almost seven million board feet of lumber to Alaska in 1890. Since it was in the timber producers' interest to keep the Alaska market, they lobbied Congress to perpetuate what Alaskans considered colonial administration.

In 1898, Congress revised the land laws to permit timber sales from public domain lands. Unfortunately, the General Land Office had neither the funds nor the people to administer such sales. A more significant law, it turned out, had already been passed: the General Revision Act of 1891 authorized the president to establish "forest reserves" on federal lands. Theodore Roosevelt, at the vanguard of the country's new conservation movement, was to use this authority with great effect in the early 1900s.

Steam-powered locomotives, such as this one belonging to the Tanana Valley Railroad, had to make stops along their routes to take on fuelwood. (Ralph McKay Collection, University of Alaska, Fairbanks)

▲This steam donkey, a steam-driven engine that pulled long cables, was used by miners in the Interior to haul ore buckets or yard timber. (Charles Bunnell Collection, University of Alaska, Fairbanks)

►Fuelwood was harvested in the winter and hauled on horse-drawn sleds. This sled is carrying between three and four cords of wood. (Charles Bunnell Collection, University of Alaska, Fairbanks)

Into the 20th Century

In 1910, Royal S. Kellogg, one of a new breed of scientific foresters, documented his examination of Alaska's forests. His report contains a terse summary entitled, "What should be done":

"The coast forests, which include most of the sawtimber of the Territory . . . have not been damaged by fire, and are but slightly reduced by cutting. They are overmature. Carefully planned cutting should take place as soon as possible. Every effort should be made to have them utilized for lumber, and especially for pulp. . . . Those of the Interior have already been seriously damaged. Their protection can not begin too soon. While the products of the coast forests need a foreign market, the interior forests with the best of treatment are not likely to supply more than a part of the home demand. If protected they will continue to furnish logs for cabins, low-grade lumber, and fuel indefinitely. Higher grade lumber required by the Interior must always be imported."

Kellogg had been sent to Alaska by the U.S. Forest Service, the agency charged with administering the newly formed Forest Reserves (soon renamed National Forests) — including the Tongass and Chugach in Alaska. One of the agency's priorities was to offer timber sales. Wood cut on these sales could legally be exported from Alaska.

In 1900, before the reserves were created, 14 sawmills reportedly cut 8.5 million board feet. In 1917, about 50 sawmills and shingle

Steam rises from the original mill of the Ketchikan Power Company, later the Ketchikan Spruce Mill, in the early 1900s. (Alaska Historical Library)

Lumber is stacked at an early sawmill at Scow Bay near Petersburg. Planked or "corduroy" roads were common during the early days in Alaska. (Lulu Fairbanks Collection, University of Alaska, Fairbanks)

mills were cutting 40 million board feet annually. Some mills, like that of the Ketchikan Power Company (later renamed Ketchikan Spruce Mill) began to export clear spruce to the states and even to England and Australia. During the century's second decade, the government sold a total of 420 million board feet in 4,000 sales from the two national forests.

Even so, Alaskans in the Interior were still importing most of their lumber. For example, it was cheaper to import construction timbers for the Alaska Railroad from Puget Sound than from southeastern Alaska. Engineers also preferred the stronger Douglas fir to spruce and hemlock.

Reconnaissance of the coastal forests in the 1920s confirmed that overall timber quality in Alaska was not as high as in the Pacific Northwest, and the amount of high-grade timber could not sustain a large lumber industry for very long. But there was a significant amount of wood suitable for pulp.

In the 1920s, B. Frank Heintzleman, a deputy forest supervisor with the U.S. Forest Service (who later became territorial governor), started his 30-year effort to get a pulp industry established in Alaska.

One early estimate predicted that the coastal forests could produce one million tons of newsprint per year on a continuing basis. In 1920, a small pulp mill was established near Juneau, but it failed two years later due to high freight costs and poor economics.

Three early-day lumberjacks pose with an immense log. Note how the end of the log has been rounded off for easier yarding. (Historical Photograph Collection, University of Alaska, Fairbanks)

A grizzly crew of lumberjacks stands with the tools of their trade. Early Alaska loggers cut timber by hand with axes and crosscut saws. (Rita Cottnair Collection, University of Alaska, Fairbanks)

In 1927, the Forest Service awarded preliminary contracts for two 8.8-billion-board-foot, 50-year pulpwood sales to Crown Zellerbach Paper Corporation and to the *San Francisco Chronicle* and *Los Angeles Times.* Both contracts were, however, cancelled in the early years of the Great Depression.

Entrepreneurs also had their eyes on forests farther north. Plans were made to harvest veneer-grade birch from the Susitna Valley, and to cut spruce and hemlock from the Chugach to feed a pulp mill. These projects were also killed by the poor economic climate of the time.

World War II brought quick change to Alaska as the nation mobilized its resources for the conflict. New military bases required lumber, some of which was supplied by Alaskan mills. Airplane-quality spruce was cut in the Tongass for the War Production Board.

After the war, concern developed over the territory's small population. A pulp industry was seen as a way to encourage permanent residents. In 1947, Congress passed the Tongass Timber Act, enabling the Forest Service to issue long-term timber contracts in spite of Native claims to parts of the forest.

One year later, Puget Sound Pulp and Timber Company and American Viscose Corporation, a rayon firm, organized the Ketchikan Pulp Company, which was awarded a preliminary 50-year contract for 8.25 billion board feet from the Tongass. The final papers were signed in 1951, and three years later the pulp mill started operating at Ward Cove near Ketchikan.

Another long-term sale, for three billion board feet, was awarded in 1954 to Pacific Northern Timber Company of Wrangell, which was to build a sawmill and pulp mill. The pulp mill was never built, however, and

the sale was reduced to a smaller amount for the sawmill.

Meanwhile, Japanese industrialists were searching for sources of timber and fiber to replace those lost at the end of World War II. After consulting with the U.S. state and defense departments, a delegation visited Alaska to investigate possibilities there. They were given a warm reception, and a few years later, a Japanese firm, Alaska Lumber and Pulp Company, was awarded a 50-year, 5-billion-board-foot contract to timber in the northern half of Tongass. Their pulp mill opened at Silver Bay near Sitka in 1959.

Two other long-term sales never got off the drawing boards: Georgia-Pacific Corporation signed a contract in 1955 for 7.5 billion board feet, mostly on Admiralty Island. They defaulted six years later. A similar sale of 8.75 billion board feet was offered by the Forest Service in 1965. St. Regis Paper Company won this award, but abandoned it two years later. The contract was then transferred to U.S. Plywood-Champion Papers, Inc. In 1970, the Sierra Club and other environmental groups sued to stop the sale, precipitating a bitter legal battle. Champion and the Forest Service cancelled the sale in 1976.

Alaska Pulp and Paper Company opened the state's first pulp mill in 1921 at Speel River, near Juneau. The mill produced up to 25 tons of wet pulp per day, which was sent to California to be made into paper. A slump in the market forced the mill to close in late 1923. (Courtesy of USFS)

By then, the annual timber harvest in Alaska was more than 500 million board feet, the rising cost of energy started changing international markets, and Alaska Natives were selecting tracts of public lands for their new corporations. The Alaska timber industry was entering a new era.

WEEKLY PRODUCTION
MOSQUITOS
THIS WEEKS OBJECTIVE
PRODUCTION TO DATE
QUOTA

◄Some of the Sitka spruce logged during World War II by the Alaska Spruce Log Program was used in the wings and nose sections of British Mosquito bombers, shown here being assembled at the deHavilland aircraft plant in Canada. Spruce was chosen because it is strong yet relatively light weight. (National Air and Space Museum, Smithsonian Institution, photo #84-10060)

▼The Alaska Spruce Log Program, established to provide lumber for the war effort, harvested about 38 million board feet of high-grade Sitka spruce between 1942 and 1944. Self-contained logging camps, such as this one at Edna Bay on Kosciusko Island, were set up in a matter of months; huge rafts of logs were towed from the camps to Puget Sound. (USFS)

The Forest Industry Today

In 1984, the first long-term pulp contract that B. Frank Heintzleman worked hard to establish in Tongass National Forest had only 20 years to go, and more than half of the contracted volume had already been harvested. Each year, more timber is cut in Alaska than was harvested during the entire second decade of the century, when Heintzleman first arrived in the territory. Yet, harvest of the Interior's forests for timber and fuelwood is probably less today than at the peak of the gold rush 80 years ago, when northern forests were cut to fuel mines, river boats and boom towns.

About 90 percent of all Alaskan wood products are exported to Pacific Rim countries, notably Japan. Our forest industry is therefore vulnerable to international financial and market cycles, as was painfully evident during the recession of the early 1980s.

In 1983 Alaska's forest industry had an output worth about $300 million, and employed some 2,800 people (down from a high of 3,600 in 1977). These figures don't include the sizable but informal subsistence economy of people cutting their own house logs and firewood.

The heart of the industry is along the coast from the Kenai Peninsula to the southern Panhandle. High-volume coastal forests are managed by federal and state governments, the Annette Island Indian Reservation and a number of private native corporations.

Timber cut from national forests must be

The pulp mill of Alaska Pulp Corporation, located on Silver Bay near Sitka, has a maximum annual processing capacity of 160 million board feet.
(Atsuo Tsunoda)

▲Federal law requires that timber harvested from national forests be processed domestically before export. Sawmills comply with the law by cutting the logs into 8½-inch-thick slabs called cants. (Walt Matell)

►A freighter loads up with cants, logs that have received the minimum local processing as required by law. Although loading continues into twilight, it takes several days to load the two to three million board feet destined for Japan. (Walt Matell)

processed by a local mill before it can be exported from the state. This is known as the primary manufacture rule, adopted in the 1920s to encourage local employment. The requirement remains in effect for spruce and hemlock logs today: logs to be exported are cut into eight-and-one-half-inch-thick slabs, called cants, before being loaded onto export ships. Smaller or defective logs that cannot be made into cants are sold to the pulp mills or chipped and then exported. When the Forest Service determines that a competitive local market does not exist for cedar, operators can receive permits allowing them to export this species as round logs.

The Alaska Division of Forestry has adopted a similar primary manufacture rule for timber harvested on state lands. A timber purchaser took them to court on this provision, and the case was appealed to the Supreme Court. No decision had been reached as of early 1984.

Private forest owners, under no restrictions, export all but their pulp and lower-grade logs in the round, as they get better prices for uncut logs.

However the wood is exported, the cost of getting logs out of the woods is high in Alaska — at least 50 percent higher than in the Pacific Northwest. So the Alaskan forest products industry has been particularly hard hit by the continued depressed timber market of the early 1980s, considered the worst since the Great Depression.

Employment By The Board Foot

The more forest products are processed, the more local jobs are created. The U.S. Forest Service has studied employment in different segments of the wood processing industry, and has found the following employment patterns:

	Workers per MMBF
Logging	2.4
Sawmilling green lumber	1.2 to 1.7
Sawmilling cants	1.0
Sawmilling dried lumber	2.0 to 2.3
Remanufacturing lumber	7.9 to 8.3
Pulp	2.4
Plywood	5.2

The figures may vary from area to area, but they do reveal general relationships.

In early 1984, the U.S. Forest Service announced that it would try to help the ailing industry by lessening the cost of harvesting timber in national forests. Measures under consideration included increasing the size of cutting units so more timber could be accessed per mile of road, lowering road construction standards, building more roads financed directly by the government and relaxing stipulations that require smaller — and often uneconomical — logs to be **yarded.**

The Tongass

Throughout the years, most of the timber harvested in Alaska has come from Tongass National Forest. In 1980, before the market slumped, 627 million board feet were harvested in Alaska; of this, 452 million board feet came from the Tongass.

About two-thirds of the annual timber cut in the Tongass comes from two long-term sales; the balance is offered as regular timber sales or special "small business set-aside" sales.

The coastal Alaska forest industry is a network of logging operators, construction firms, transport companies, mills, service businesses, private forest landowners and government agencies. They operate in a semi-remote environment, and have learned to

Forest workers walk along a makeshift trail through a future timber road right-of-way.
(Caribou Trails Photography)

◄**A tractor builds a road to access more timber at the state's Icy Cape #1 timber sale, near Cape Yakataga.** (Rick Rogers)

▼**Crushed rock for road construction is quarried as close to roads as possible to avoid expensive hauling. At up to $200,000 per mile, road construction makes logging in southeastern Alaska quite expensive.** (Caribou Trails Photography)

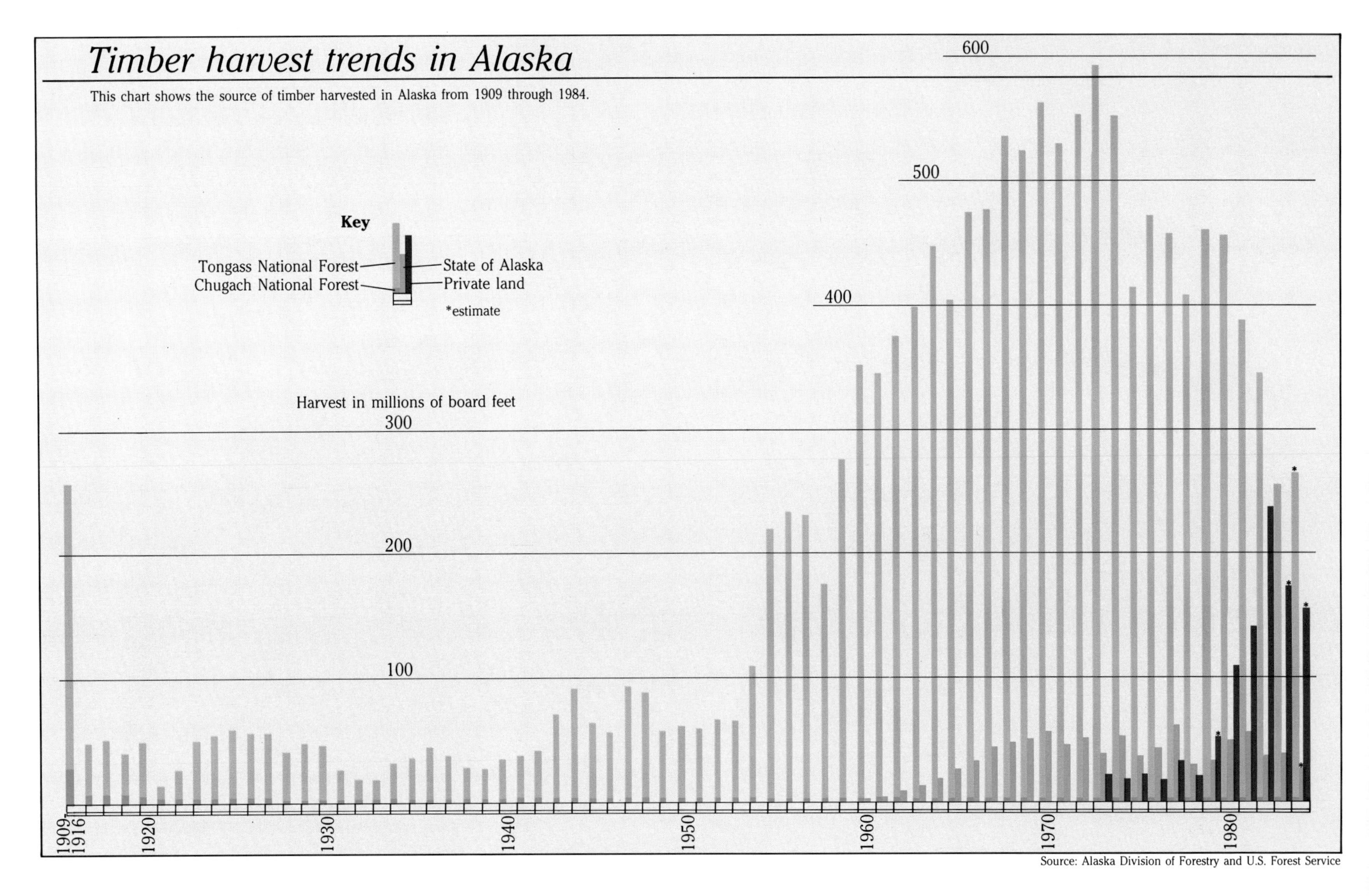

cope with one another: loggers, for example, have been known to grumble more about government paperwork than about devil's club or howling storms. On the other hand, government foresters have been mandated to manage public timberlands for a variety of uses in addition to timber.

Although more forests in Alaska remain public, ownership has changed in recent years, as reflected in the table below.

Harvesting Interior Forests

The annual timber harvest in interior Alaska is estimated at less than 15 million board feet, from state lands, native allotments and private corporation lands.

Now that landownership of vast areas is no longer in limbo, industry can start to plan with a better idea of who owns what. Unfortunately, the new property boundaries are often geometric, with little respect for

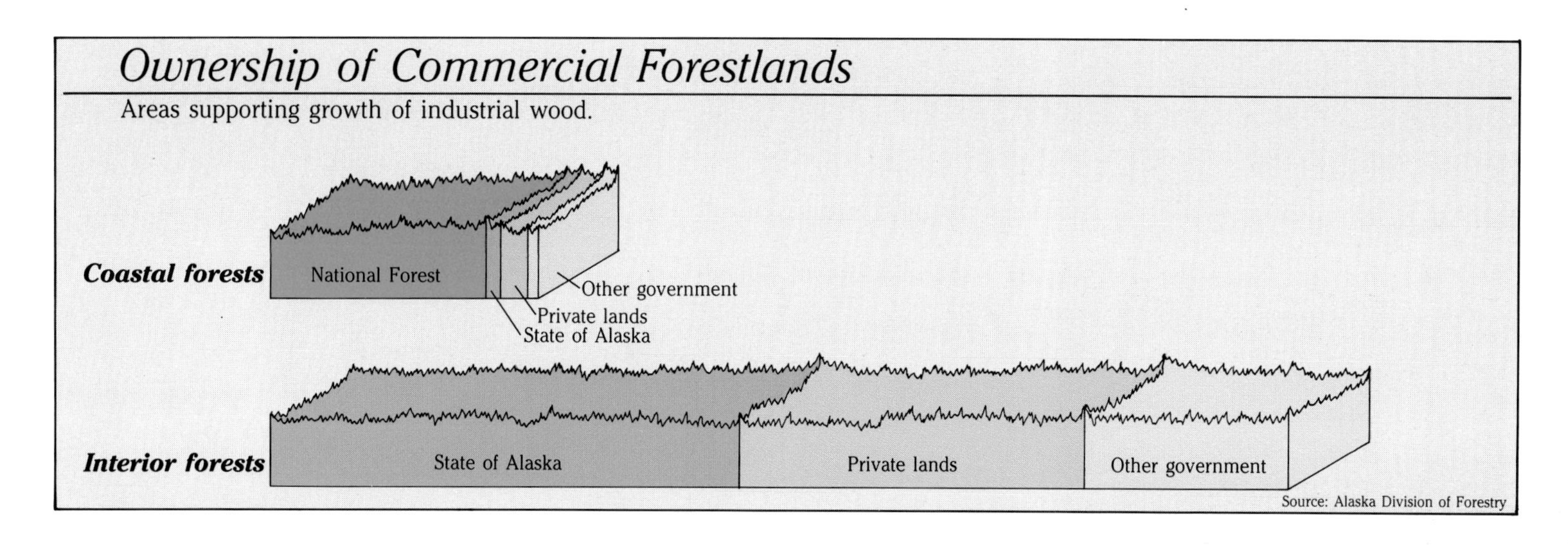

ecological or topographic breaks. Across these man-made divides, managing natural resources like forests will be a challenge.

Interior forests are a mosaic of stands of varying types and densities. Long **stringers** of better sawtimber usually follow river valleys. Such linear stands are more costly to harvest, because loggers must build more roads per acre than in forests where they could use spokelike road systems. Some planners think that frozen rivers can become logging roads; others foresee large timber-processing complexes at the lower ends of the Yukon and Kuskokwim rivers, fed by logs floated or barged downstream during summer.

Most logging in the Interior occurs on state lands. Trees are felled by chain saw or powerful tractor-mounted shears, and hauled by truck to local mills, where they are cut into green rough lumber, house logs or specialty products. Timber sales are typically less than two million board feet each, and have been offered on lands the state was fairly sure it wanted to continue managing as timberland.

The largest timber operation in the Interior in recent years was on private forestlands: from 1975 to 1981, Toghotthele Corporation harvested at least 16 million board feet of

Logging in the Interior often waits for winter. Here, Mary Shields skids a log with the help of her dog team. (Dean Argyle, USFS)

▲**A felled spruce shows the "hinge" left by the timber faller. Since the tree falls perpendicularly to this hinge, a good cutter can lay a tree with extraordinary precision.** (Caribou Trails Photography)

►**From a springboard high above the ground, a timber feller watches as a large spruce lifts off its stump. Springboards such as this are used to cut above the butt swell, or flared area, at the base of the tree.** (Hank Nelson, reprinted from *ALASKA*® magazine)

Long-Term Sales

Although more had been offered, only three long-term timber offerings became actual sales in the Tongass. Events leading up to the sales are discussed in the previous chapter.

As of early 1985, Louisiana-Pacific Corporation owned the Ketchikan Pulp Company contract. The Pacific Northern Timber contract, which recently expired, was owned by Alaska Pulp Corporation (formerly Alaska Lumber and Pulp Company), whose parent company is Alaska Pulp Company, Ltd., of Tokyo. Below are the dates and volumes involved:

	Ketchikan Pulp Co. (now LPK)	Alaska Pulp Corp. (ALP)	Pacific Northern Timber
Effective dates	1951-2004	1956-2011	1954-1982
Total volume, MMBF	8,250.0	4,974.7	693.1
Annual release volume	192.5 MMBF	104.2 MMBF	24.8 MMBF

The "annual release volume" is the average volume the Forest Service offers for harvest per year to the companies.

Since markets fluctuate, the companies may wish to cut more or less any given year. The contracts spell out certain arrangements that allow them to harvest some backlog volume not cut in prior years. Contracts differ on the types of trees to be cut: Alaska Pulp's contract specifies certain percentages of spruce, hemlock and cedar, while LPK's contract has no species limits.

Both remaining long-term contracts are subdivided into five-year periods, for which harvest areas are specified and stumpage rates are set. (**Stumpage** is the price purchasers pay for trees they cut. Normally, federal timber is auctioned to the highest bidder above set minimum rates. Since each long-term sale has only one buyer, however, stumpage for these sales is calculated by the government.) The Forest Service has developed stumpage formulas that take into consideration things like logging costs, costs of building roads, the selling prices of end products from timber to be be released and a factor for profit and risk.

When timber markets are poor, stumpage is lowered. For example, LPK's rates in 1983 were close to the minimums; the company was paying $2.87 per MBF for spruce, $1.97 per MBF for hemlock, $1.62 per MBF for redcedar and $13.42 per MBF for Alaska-cedar. The company bears the cost of road construction in the harvest area.

The Fourth of July Logging Carnival in Ketchikan provides a mid-season break as loggers come to town to celebrate and to participate in events testing skills honed in the woods. (Walt Matell)

Some logging camps in southeastern Alaska are floating villages, complete with homes, office and school. The camps are towed periodically to current logging locations. Camp children, such as the group above at Whale Pass, are required to wear life preservers at all times.
(Above, Walt Matell; right, USFS)

White spruce logs from Toghotthele Corporation lands near Nenana are loaded onto a truck for export in the late 1970s. (Dean Argyle, USFS)

white spruce near Nenana, and shipped it on the Alaska Railroad to the coast for export.

Many communities in the Bush have small portable mills, with which they cut subsistence quantities of local timber. Some villages are near good-quality stands, and others are not. Even stands considered noncommercial (annual growth less than 20 cubic feet per acre) may still be adequate to keep small communities supplied with house logs, lumber and fuelwood. Communities considering development of their timber resources can receive forestry assistance from various state and federal agencies.

Alaska's Sawmills and Pulp Mills

Although sawmills operate in most communities in the state, most milling is done at a dozen mills capable of processing more than 10 million board feet per year. These are all in the coastal region, close to the big timber. During low markets, some may stop operating at full capacity or shut down.

Small local firms produce lumber and house logs on site. Many use portable circular-saw "Volkswagen" mills (so-called because they are often powered by VW engines), which typically produce less than one million board feet per year.

Marketing Alaska's Forest Products

In general, Alaska log and cant sellers are the last to be competitive in a rising timber

Although most building materials used in southeastern Alaska are imported from the Pacific Northwest, a few small businesses are marketing locally. In Craig, Bruce Brown makes redcedar roofing shakes from local wood with this spinning taper saw. (Walt Matell)

market, and the first to become non-competitive in a falling market. So they must target their markets carefully.

Japan remains the largest single market, consuming better than 90 percent of Alaska's timber exports. Imports from Alaska, however, constitute less than three percent of the wood that Japan imports from all sources. So from the buyer's perspective, Alaska is not very important in terms of quantity. What the Japanese like about Alaskan wood is its quality — the big, clear-grained logs that are becoming more scarce from other sources. Sometimes called green gold, select quality spruce and **peeler** hemlock (suitable for making plywood) are more profitable to harvest than lesser grades. Not surprisingly, industry would prefer to cut better grades, bigger logs and more spruce than the present mix.

In areas considered commercial forest in the Tongass, the species mix is about 28 percent Sitka spruce, 64 percent western

The Kuskokwim Corporation

The Kuskokwim Corporation, a merger of 10 village corporations along the Kuskokwim River, owns almost one million acres of mostly white spruce and birch forest along the river between Lower Kalskag and Stony River.

A study to determine the development feasibility of the forests along the middle Kuskokwim River concluded that 109,000 acres had commercial potential, and that this land could sustain an annual harvest of 1.6 million cubic feet of spruce per year. Three development options were presented: to export the spruce as round logs; to set up a 7.5-MMBF-per-year sawmill producing green lumber for export; or to establish a smaller 1.7-MMBF-per-year mill along the river to supply local markets.

Due to the high cost of logging this remote area, the short working season, the limited local market and the depressed state of the export market, none of the three options is considered economical today. But The Kuskokwim Corporation is optimistic that economic conditions may change during the next decade, and if so, that it can gradually enter the timber market when conditions improve.

Timber Processing Capacity in Alaska

This list includes only mills with a rated capacity of 5 MMBF or more per year; actual production depends upon many factors, especially the economy. There are an estimated 400 other mills with smaller capacities throughout the state.

Willow

Susitna Mill & Log, Inc.
Sawmill
5 MMBF

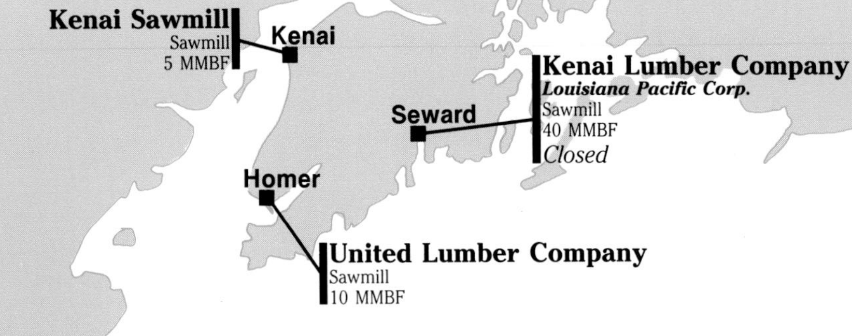

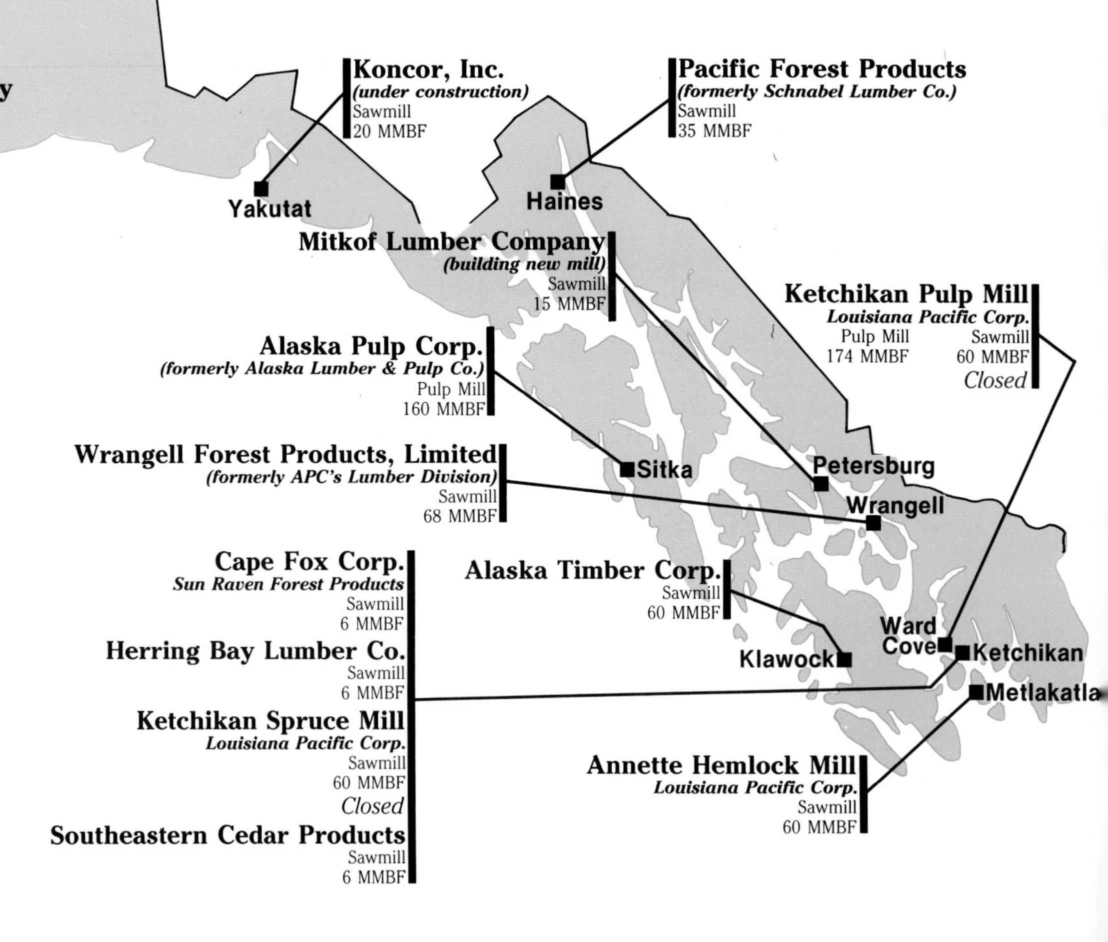

Source: Alaska Department of Natural Resources "Alaska Forest Industry Directory, 1983-84"

This 1961 photo shows the community of Hollis, on the east coast of Prince of Wales Island, site of the state's first large-scale pulp logging. (USFS)

Sawdust and wood chips, by-products of all sawmills, can be pulped or used as an energy source. (Fjord Photography)

A log scaler estimates the grade and number of boards that can be sawn from this log. Scalers rely on formula tables, lots of experience, and tools such as this scale stick, which is used to probe for defects which lower the log's grade. (Walt Matell)

Round logs imported from Alaska are sawn into house posts and other construction lumber at the Asama Shokai Company mill at Wakayama, southwest of Tokyo. Japan has more than 10,000 sawmills of varying sizes processing timber from Alaska. (Both by Tom Asakawa, Alaska State Asian Office)

Native corporations like Shaan-Seet, Inc., of Craig are harvesting their lands for export timber. Their Unit 1 is evident as a large clear-cut behind the town of Craig. (USFS)

hemlock and 8 percent cedar and others. The proportion of spruce is expected to increase in the second growth stands that are replacing old growth now being cut. But in the planned 80- to 120-year rotation, the second growth stands probably will not produce logs as fine-grained or as large as those available from old growth stands today.

Buyers of Alaska wood pay more for round logs than for cants: in 1982, for example, the average price per thousand board feet of export softwood lumber was $336, while round logs were going for $478. So in spite of the downward turn of the market for cants, the round log export market has been steady or rising.

In the past, if the Japanese wanted Alaskan timber, they had to settle for cants, since practically all of the timber came from national forests. Indeed, some mills in Japan have been specifically designed to process cants. But now they can also buy round logs, which offer them more flexibility and greater efficiency in milling for their discriminating customers.

The Alaska timber industry is looking around to diversify into other markets. Unfortunately, the rest of the United States is not a good prospect except for specialty items. Not only are timber-rich Canada and the Pacific Northwest closer to this market, but the Jones Act restricts interstate shipping to vessels registered in the United States, driving transporation costs out of sight. So

The Japanese Timber Market

When a shipment of logs or cants imported from Alaska arrives in Japan, it is immediately sorted and graded again. Japanese mills are extremely quality-conscious about the timber they will saw, and cut with greater attention to grain alignment than most of their counterparts in the United States.

Ninety-one percent of imported Alaskan timber ends up in Japanese houses, frequently as exposed beams and posts. Such timbers are custom-cut at mills before being sent to the building site. Grain orientation and lack of defects are very important. Blemishes acceptable in the American "cover it with paint or sheetrock" construction method are not tolerated in traditional Japanese carpentry.

But, due to the expense, building techniques are slowly changing in Japan. Some contractors are experimenting with veneers or imitation vertical-grain laminates, and even western-style stud and plywood construction. Newer, less expensive houses use less wood than traditional houses. If these trends continue, the demand for high-quality Alaskan timber may decrease.

The remaining 9 percent of Alaskan timber is used for specialty items like musical instruments. The Yamaha piano, for example, is made from Sitka spruce. Alaska-cedar is a highly prized substitute for *hinoki* or Port Orford cedar, used in furniture.

The Japanese also have forests of their own, and are managing them intensively. Their 1980 domestic harvest was estimated at between 30 and 45 million cubic meters (6,500 to 10,000 MMBF, or 10 to 15 times the annual Alaska cut), a volume they plan to double by the turn of the century.

Sealaska shareholder Dallas Chaney is learning the timber business, working as a junior scaler at the corporation's sort yard near Klawock. (Larry McNeil, Sealaska)

Sealaska Timber Corporation

The largest private landowner in southeastern Alaska, Sealaska Corporation had acquired title to 201,000 acres of former national forest lands by the end of 1983, and expects to receive about 130,000 more acres in the next few years.

Much of their property is prime timberland. Sealaska formed a subsidiary, Sealaska Timber Corporation, to develop and market their timber resources and those of some of the other Native corporations in southeastern Alaska as well. Logging started in 1980, and is now centered on Prince of Wales and Dall islands in the southern Panhandle.

In 1982, Sealaska's resources put the company on the "Fortune 1000" list. In the four years since 1980, Sealaska Timber Corporation has marketed more than 440 million board feet of logs, exporting most to the Orient. Sealaska will undoubtedly be a major force in the future of Alaska's forest industry.

◄Round logs from several native corporations are sorted at Sealaska Timber Corporation's dock near Klawock in preparation for export to the Orient. (Walt Matell)

▼After sorting by species and grade, bundles of logs are put into the water for storage or transport at the Sealaska Timber Corporation sorting yard near Klawock. (Walt Matell)

Redcedar logs from Cape Fox Corporation lands are stacked awaiting export. Note the inventory control tags that can be scanned like cans of peas in the supermarket. (Walt Matell)

Alaskans are seeking to develop trade relations with other countries of the Pacific Rim.

One of these is the People's Republic of China. With a population of one billion and a forest base only one-tenth of the United States' on a per capita basis, China is considered a huge potential market for Alaskan wood products, particularly because it will also buy the lower grades. In 1983, a trade mission led by Alaska senator Frank Murkowski visited China to get acquainted with potential trade partners in the various ministries there. The Chinese have slowly been getting acquainted with Alaskans and their wood by buying small quantities of various products from different suppliers. This market will take time and patience to develop.

It is important, also, that work be done to develop one market closer to home — the market in Alaska. One hundred million board feet of lumber is imported annually into the railbelt area of Alaska, an amount equal to more than 15 percent of the state's yearly harvest.

This is a problem that has been apparent for a long time, but has yet to be resolved. Why indeed are Douglas fir timbers from Oregon used to build log houses on the Yukon River? Why are there so few dealers offering planed, kiln-dried and graded domestic lumber in Alaska?

Richard Evans, President of United Lumber Company in Anchorage (one firm that kiln-dries lumber) offered this answer: "I can buy a two-by-four [from the Pacific Northwest] cheaper than it costs me to produce it locally. The mills down south get enough returns from other parts of the log that they can afford to dump the two-bys on us." Others in the business agree with Evans, and maintain that the only thing preventing them from entering the domestic finished lumber market is that they would go broke. That is hardly an incentive to answer the call for a domestic lumber market.

Since dried and graded lumber is not economical to manufacture in Alaska, could the Interior trees be chipped into particleboard instead? Fairbanks Industrial Development Corporation and the Forest Service studied the proposal, and determined that such a facility, producing 50 million square feet (1.5 million 4' x 8' sheets) per year, would quickly saturate the Alaskan market and would require outside markets as well. Moreover, the cost per sheet of this Alaskan particleboard was found to be higher than either imported particleboard or plywood.

In Alaska, mills that do manufacture lumber cut it rough and air-dry it at best. Unfortunately, most building codes specify graded lumber, so contractors are forced to use imported lumber. Do-it-yourselfers and contractors doing rough-wood projects like docks and outbuildings will buy local lumber, but only it it costs less than imported wood.

Building contractors often find that using imported building materials is more economical than purchasing locally. This house in Nulato, on the Yukon River, was constructed of squared Douglas fir logs imported from the Pacific Northwest. (Jim Anscomb, Architects GDM)

Most mills in Alaska air-dry their green lumber. This stack of two-by-sixes is drying at Northland Wood Products in Fairbanks. (Walt Matell)

Paul McArthur works at the strenuous task of peeling logs to be used to build the Cooper Landing Library. (Tom Walker)

Tom Walker cuts a scribe line on a log for the home he is building. Log homes have become increasingly popular in recent years. (Tom Walker)

House logs tell a more encouraging story. Log homes and even log commercial buildings are gaining in popularity. Many firms specializing in turning or slabbing logs, or even precutting logs for entire buildings, have been established in recent years. A favorite tree is white spruce, which, if skillfully used, makes beautiful and durable log buildings. People who decide to build with logs often do so because they want an Alaskan-style building, even though it may be more expensive than a conventional frame-lumber structure. So the house log market is somewhat insulated from Pacific Northwest and Canadian lumber competition. But not to the extent that house-log sellers will capture the entire Alaskan housing market.

If technological improvements can make logging and small mills more efficient, then perhaps the costs of producing finished timber products for local markets will fall to a level these markets can afford. The Forest Service, Cooperative Extension Service and Bureau of Indian Affairs offer technical workshops and efficiency evaluations to interested operators, but suggested improvements usually require additional investments, and therefore capital. Few investors are willing to lend money when the markets are as low as they have been recently.

The state has enacted a law which requires that "in a project financed by state money in which the use of timber, lumber and manufactured timber products is required, only [such] products originating in this state from local forests shall be used whenever

◄**In Alaska, there are more than 400 small portable "Volkswagen" type sawmills, such as this one in Southeast.** (Walt Matell)

▼**There are many small sawmills on the Kenai Peninsula, such as the mill operated by Abe Abrahamson on the south fork of the Anchor River. Here, Abrahamson gauges and sets the width of the slabs.** (Janet Klein)

practicable." The Achilles' heel of this law, of course, are the last two words.

The State Legislature provided a more substantial boost to the forest industry in 1983 by designating the Tanana Valley State Forest. Its 1.6 million acres include tracts along the Tanana River from north of Denali National Park and Preserve to Tok. These state lands are to be managed for multiple use of their forest resources, including timber harvest. A year earlier, the state had established the first state forest, which included a bald eagle preserve, in the Chilkat Valley north of Haines, thus settling a long-standing conflict between timber and wildlife resources of that forest.

Although the state had been offering timber sales for many years, there had always been uncertainty about whether or not more

Louisiana-Pacific

◄Depending on the timber market, at times logs are imported from Canada to the Ketchikan Pulp Mill. Here, a self-dumping barge unloads 1.5 million board feet of Canadian pulp logs. (Walt Matell)

Logs destined for the Ketchikan Pulp Mill are lifted out of the water by a huge bundle crane. (Walt Matell)

At the Ketchikan Pulp Mill, bleacher-man Torleif Dale (left) takes a sample of pulp from one of the washers that remove impurities and treat pulp with chemicals. Below, the continuously running pulp machine produces a 20-ton jumbo roll of pulp in about 40 minutes. The high-grade pulp is exported to the Lower 48 and countries around the world, where it is redissolved and made into rayon, cellophane, and even a fiber food additive. (Both by Walt Matell)

timber would be available in the future from state lands and, indeed, whether these forestlands would remain in state ownership. That cloud of doubt was lifted, and hopes have kindled for additional state forests in other parts of Alaska.

The Forest Industry In 21st Century Alaska

Alaska's forest industry is closely tied to the international wood and fiber marketplace. It will probably remain so for many years, because the state's timber resource will continue to exceed local demand, even if the domestic market is fully developed.

Short-term predictions show increasing worldwide demand for wood products.

Many former public forests are now private property, bringing a change to Alaska's timber industry. (Walt Matell)

Whether or not Alaska timber will help meet this demand is a key question for the industry. Will Alaska's contribution continue to be large, fine-grained logs and cants and high-grade dissolving pulp? Or will it be able to move into lower-grade markets as well? This is especially important for the industry in interior Alaska, but it is also a consideration on the coast. Available old growth stands there are gradually being replaced by second growth stands, which are more productive and faster-growing but will never grow large, high-quality trees if harvested on a rotation schedule that maximizes growth. The industry that harvests this second growth will be different from today's industry, relying more on smaller trees with larger growth rings.

Longer-term forecasts vary, and depend on many economic and social factors, including the worldwide economy, the price of energy and fiber substitutes and how intensively Alaska's forests are managed. The future will also be influenced by nonhuman factors like insect and disease epidemics, productivity of forest soils, shifts in climatic patterns, and whether the forests maintain sufficient genetic diversity to cope with possible environmental changes.

Like the forest resource itself, the forest industry will probably continue to go through cycles and may change with time. But with reasonable care and foresight, it can be as perpetual and renewable as the forests.

Energy From Alaska's Forests

Alaskans have relied on their forests to provide firewood for thousands of years. By the middle of this century, cheap petroleum replaced wood as the dominant fuel for heat and power in all but the most remote areas. As oil prices have risen in the last 10 years, however, the pendulum has swung back, and demand for fuelwood has taken off like wildfire.

So great is the demand that foresters are now planning to manage certain forests primarily for fuelwood. And those with an eye for the future are contemplating hybrid or exotic strains of quick-growing trees or shrubs to be used in energy plantations. This is an accurate phrase, since burning wood actually releases solar energy a plant has collected and stored. If one imagines forests as being huge solar energy sponges, then the amount of fiber, or **biomass**, in the forest is a good indicator of how much energy has been stored.

The U.S. Forest Service estimates the biomass of the aboveground portion of trees in Alaska to be at least three billion tons of green fiber. When air-dried, this fiber has the energy equivalent of about 24 quadrillion **Btus** (British thermal units). That is equal to about 2,000 times the annual electrical power production in Alaska, or the energy potential of one-half of the estimated recoverable oil reserves at Prudhoe Bay.

Total forest biomass is significantly higher than this, however, as much of it is in the

Steve Dunphy and Don Albrecht of Don's Firewood salvage logs from a two-year-old state clear-cut in the Tanana Valley. Working at a steady pace, the commercial firewood operation can collect about five cords a day. (Walt Matell)

non-tree vegetation of the forest floor or below ground. This is particularly true in black spruce forests, where the thick, spongy layer of mosses and forest debris blanketing the ground may constitute more than 75 percent of the total biomass. Scientists jokingly refer to them as stands of feather moss with a black spruce **overstory.**

Putting up firewood for the winter is a way of life throughout Alaska. Wood is usually plentiful for this subsistence use; many people literally cut firewood in their backyards. Now city dwellers, too, are installing wood stoves. In timber-rich southeastern Alaska, most suitable firewood along the few roads out of the larger towns has already been picked up by suburban loggers. Cutting logs salvaged from beaches is one way around this, as is buying cordwood from entrepreneurs. Commercial woodcutters are well established in Anchorage and Fairbanks, but many people still like the experience of cutting their own wood — and their fuel bill.

State officials estimate that about 90,000 cords of firewood are harvested annually, and that the demand at the turn of the century may increase to more than 700,000 cords. That's a lot of wood, equal to about 50 million cubic feet. If harvested from hypothetical forests growing exactly 20 cubic feet per acre per year (the minimum potential growth to classify a forest as timberland), it would take 2.5 million acres of such forests to provide this volume of wood on a sustained basis.

Thousands of acres of forest near Delta Junction have been cleared and burned to make way for barley crops. But the debris left after clearing, shown above, contains a significant amount of biomass energy, which could be used to heat people's homes. (Left, Gary Michaelson; above, George Sampson, USFS)

Forester Butch Ruppert holds two fuel logs made from land-clearing debris. The black log, made mostly from compressed moss, yields about 6,500 Btu per pound; the other log is made from black spruce chips (shown above right) and contains about 8,000 Btu per pound, an amount comparable to supermarket fuel logs. (Both by Walt Matell)

Many forests, of course, grow a lot faster than this, so the acreage required would actually be somewhat less.

In the old days, tepee burners stood beside all sawmills, burning scraps and residues. Today, the Alaska Timber Corporation mill at Klawock burns its waste to generate power, and perhaps other mills will install similar systems.

The Alaska Village Electric Cooperative has studied the possibility of using digesters to turn wood chips into a combustible gas to fuel diesel generators that provide electrical power in remote communities. Preliminary cost evaluations showed that using the digesters would be more expensive than continuing to use diesel fuel; however, that situation is likely to change if the price of oil increases again.

The Fairbanks Municipal Utilities System is planning test burns of wood chips mixed with coal in its electrical power plant. It and researchers from the U.S. Forest Service and University of Alaska are trying to find out whether it would be economically and technically feasible to produce wood-generated electricity for Fairbanks.

Even if more expensive, one obvious advantage of using wood for power is that, unlike diesel fuel and coal, wood is a renewable resource. A bonus for Fairbanks might be less air pollution; tests in other areas have indicated reduced particulate emissions when wood was added to coal fuel.

The renewable nature of forests makes using them as a source of energy appealing. If done carefully with full attention paid to side-effects, biomass conversion may become a large industry. Such use of Alaska's forests will involve technology more sophisticated than the pioneer's ax and maul, to be sure, but the biological principles underlying our children's use of the forest will be exactly the same.

◀Smoke rises from a tepee burner, used to dispose of scraps and other debris, at a sawmill in Haines in 1975. Such burners are no longer used. (Charles Kay)

▼Because of a lack of access roads, firewood is often hard to find near some of the larger communities in southeastern Alaska. To help alleviate this problem near Ketchikan, the Forest Service designed a timber sale that specifically required lower-grade logs to be yarded for local woodcutters. (Carl Holguin)

If a wood biomass chip industry is to be successful, it needs a steady supply of the chips, meaning that they would have to be stockpiled. Forest service researchers in Fairbanks are monitoring this chip pile, which has been imbedded with temperature sensors, to find out how long-term storage affects the quality of the chips.
(Walt Matell)

A methane gas mixture is produced from wood chips by this prototype gasifier. Such machines could one day fuel electrical generators in remote villages, converting biomass to kilowatts.
(Ken Kilborn, USFS)

Alaska Timber Corporation vice president Gregory Head is dwarfed by the mill's new electrical power generating plant at Klawock. The plant, which will run on wood wastes, will eventually have a total capacity of more than four megawatts. (Walt Matell)

Fire and Alaska's Forests

Although much of it sits on permafrost, the black spruce taiga of interior Alaska is second only to the notorious chaparral of southern California as the most fire-prone forest in the United States. In interior Alaska, the forests burn often, and there is a lot to burn.

Large Fires in Alaska

More than half a million acres burn in Alaska each year; this average includes years in which only a few thousand acres burned, as well as the big fire years of 1957, 1969 and 1977, when a total of more than 11 million acres were consumed. Scientists estimate that about two million acres burned each year before modern fire detection and suppression methods were introduced in the last few decades. [See chart on page 158.]

Today, about 65 percent of all fires are caused by people. But since they are usually close to settlements and highways and are therefore controlled more quickly, such fires account for less than one quarter of the acreage burned.

Controlling wildfires is the shared responsibility of three government agencies. Generally, the Bureau of Land Management handles the remote northern and western sections of the state; the State of Alaska is gradually taking over from the BLM in southcentral Alaska and along transporation corridors of the Interior; and the U.S. Forest Service has responsibility for southeastern

Tremendous amounts of energy are released as fires spread to the crowns of trees, where fuel and oxygen can mix well. (Tom Evans, BLM)

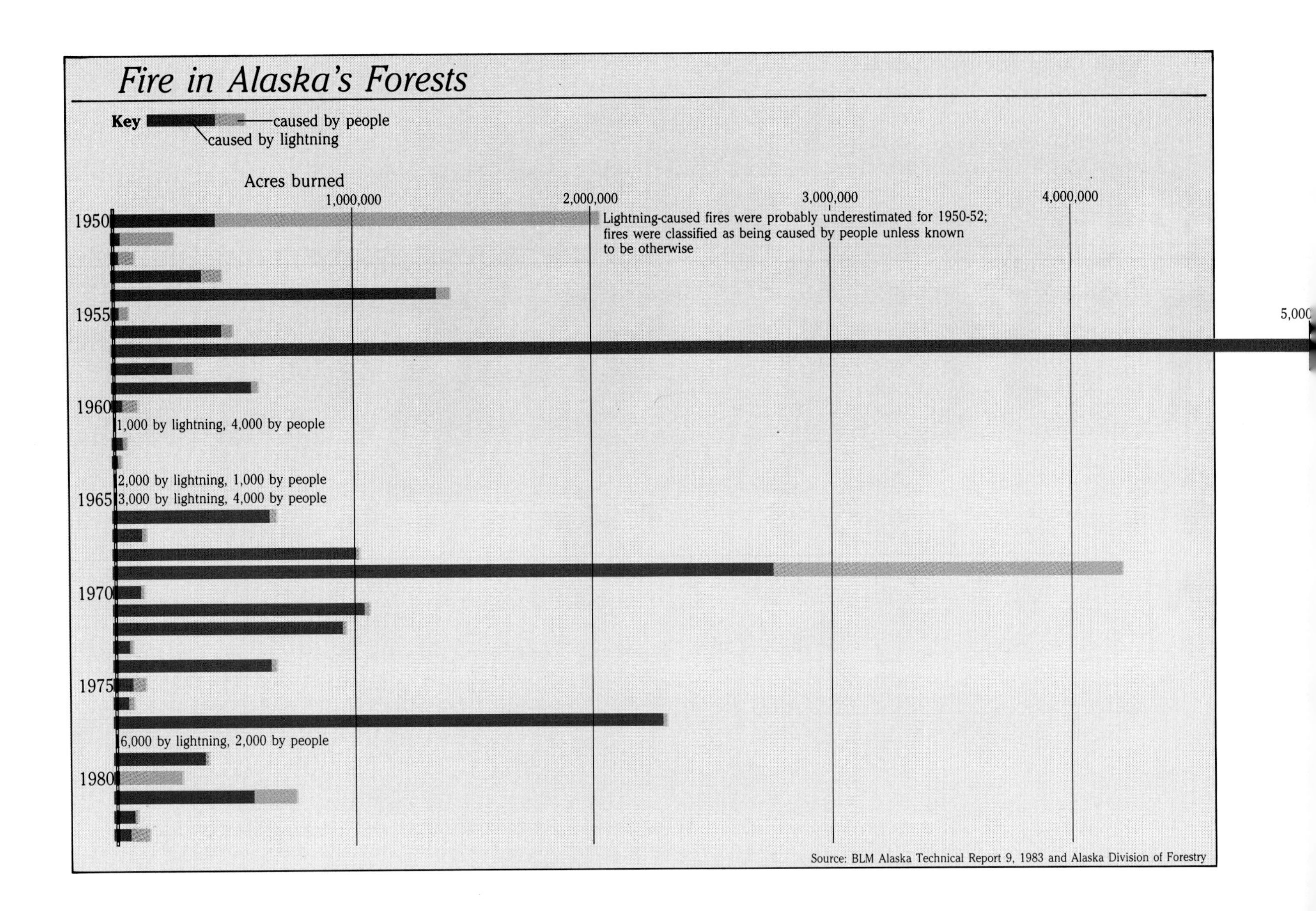
Fire in Alaska's Forests
Key
caused by people
caused by lightning
Acres burned
1,000,000
2,000,000
3,000,000
4,000,000
5,000
1950
1955
1960
1965
1970
1975
1980
Lightning-caused fires were probably underestimated for 1950-52; fires were classified as being caused by people unless known to be otherwise
1,000 by lightning, 4,000 by people
2,000 by lightning, 1,000 by people
3,000 by lightning, 4,000 by people
6,000 by lightning, 2,000 by people
Source: BLM Alaska Technical Report 9, 1983 and Alaska Division of Forestry

This photo shows two ways of delivering water to a fire: by helicopter air-drop or by "bladder-splatter" backpack. (BLM)

Alaska and Chugach National Forest. Under the Alaska Native Claims Settlement Act, the federal government is responsible for fire suppression on native lands.

Providing wildfire protection in Alaska is a big job, especially in hot, dry years when extra personnel and equipment may have to be imported from Outside. The task will increase in the future, as government land disposals encourage more people to move to remote areas to live and work; these people will, no doubt, demand protection for their new communities.

Not only will this cost more, but researchers are concerned that total fire suppression will alter long-term ecological cycles. In the Interior, wildfires are regular occurrences: most areas of the taiga have been burned at least once in the last 250 years. The taiga doesn't survive in *spite* of these fires; it survives *because* it is burned periodically. Therefore, foresters now talk in terms of fire management, rather than fire fighting, implying that not all wildfires should be attacked automatically.

Resource managers recognize that there are no easy answers when dealing with wildfire. Decisions they make today are not so simple as "let them all burn," or — as was the case until very recently — "attack all fires and keep 'em small." Instead, they are formulating a range of possible responses before dry weather and lightning put them into the yearly fire season.

The Phoenix Forest

On the cold taiga, layers of organic material accumulate to great thicknesses on the forest floor, insulating the soil from the summer's heat. This thickens the permafrost layer, and slows decomposition that would otherwise recycle nutrients. Most tree species in the Interior need mineral soil in which to germinate their seed. Seeds that land in the spongy organic layer usually do not survive because the layer cannot hold moisture during hot summer days.

The dried-out organic matter also happens to be perfect tinder, an open invitation for wildfire. If the inevitable fire succeeds in removing some of this layer, mineral soil is exposed to summer warmth, the permafrost zone lowers, and nutrients are released. The site becomes more productive as conditions start favoring seedlings and new growth from underground roots.

The type of forest that returns depends on what was there originally, and on the severity of the burn. For this reason, the taiga is an ever-changing mosaic of different types and ages of forests.

▲Smoke jumpers arrive at a fire in the Interior in new, gliderlike parachutes. Aerial transport has made fire fighting more efficient, and allows crews to reach previously inaccessible fires quickly.
(Jeff Bass, BLM)

►Fire fighters work to backfire an area near an advancing wildfire. The technique burns away vegetation, depriving the wildfire of fuel.
(Tom Evans, BLM)

After a fire has been contained, the mopping up operation is hot and tedious. (BLM)

Fire fighters retrieve their parachutes after containing a 65,000-acre fire near Beaver in the Interior. Cost of suppressing the 1977 fire was $369,000. (Alaska Division of Forestry)

The Tanana-Minchumina Plan

The task seems complex: take a 48,000-square-mile area in the heart of Alaska, encompassing remote communities like McGrath, Ruby, Tanana, Manley Hot Springs and Rampart, as well as railbelt towns like Nenana and Fairbanks. Throw in Mount McKinley, the upper Kuskokwim River and parts of the Yukon and Tanana rivers. To this broad mosaic of forests and tundra, overlay the checkerboard of various landownerships. Now figure out one plan that will guide wildfire management in the area.

That is what the Tanana-Minchumina Interagency Fire Management Plan accomplished.

A planning team was assembled from representatives of native organizations and government agencies. They compiled information on the resources and fire occurrences in the unit and the goals of the land managers. They also held public meetings in communities within the unit. The final product was a set of maps which assigned each area within the unit one of four basic options:

Critical protection areas receive top priority and immediate and aggressive fire suppression. Typically, these sites are around communities and areas where human life and property might be endangered by wildfires.

Full protection areas also get immediate and aggressive fire control, but at a lower priority. Examples include historical sites, areas with high resource values, and lands requiring protection similar to that provided in the past.

In *limited action* areas, the land managers have decided in advance that they want to maintain a natural fire cycle, or that the resources there do not justify the expense of fire suppression. Wildfires are allowed to burn within these areas, but action is taken to prevent fires from straying onto adjacent areas with higher protection.

In *modified action* areas, land managers have some flexibility between full protection and limited action.

The options are reviewed yearly, and refinements are made based on accumulated practical experience. The objective is to let everyone know in advance what level of protection a particular piece of land will get.

Fire Management Plans

As federal land was carved up into new ownerships, land managers realized that they would have to coordinate their fire planning efforts. They formed the Alaska Interagency Fire Management Council, a working group with representatives from various government agencies and private landowners.

The council has coordinated fire management plans for five areas, stretching across the state from the Kuskokwim to the Copper River.

Wildlife and W[illegible]

The Smokey Bear comics we all read as kids showed families of animals fleeing in terror from approaching forest fires. According to forest ecologists, Smokey's implied message that wildfires are always bad for wildlife needs updating. A hot, fast-spreading wildfire can indeed kill animals. But more importantly, fires rejuvenate habitats necessary for long-term wildlife survival.

One of the best-known examples of this in Alaska is the dramatic increase in moose populations following fires. The early succession of aspen, willow and other browse species follows quickly after fire, and the moose love it. Maximum browse is typically available between five and twenty years after burning; sometimes the moose themselves prolong the browse period by eating enough of the new vegetation to delay growth above their reach. Moose apparently depend on a

Dale Taylor, ecologist with the Bureau of Land Management, shows areas of the state covered by fire-management plans in 1983. Large units are broken down into smaller areas, each of which is assigned one of four levels of fire protection. (Walt Matell)

This printout from the Alaska Fire Service's lightning detection system shows 705 strikes, in a swath across northeastern Alaska and in a cluster on the Seward Peninsula. By combining this information with data from a network of automatic weather stations across the state, fire managers can make informed decisions about where to locate their crews. (Courtesy of BLM)

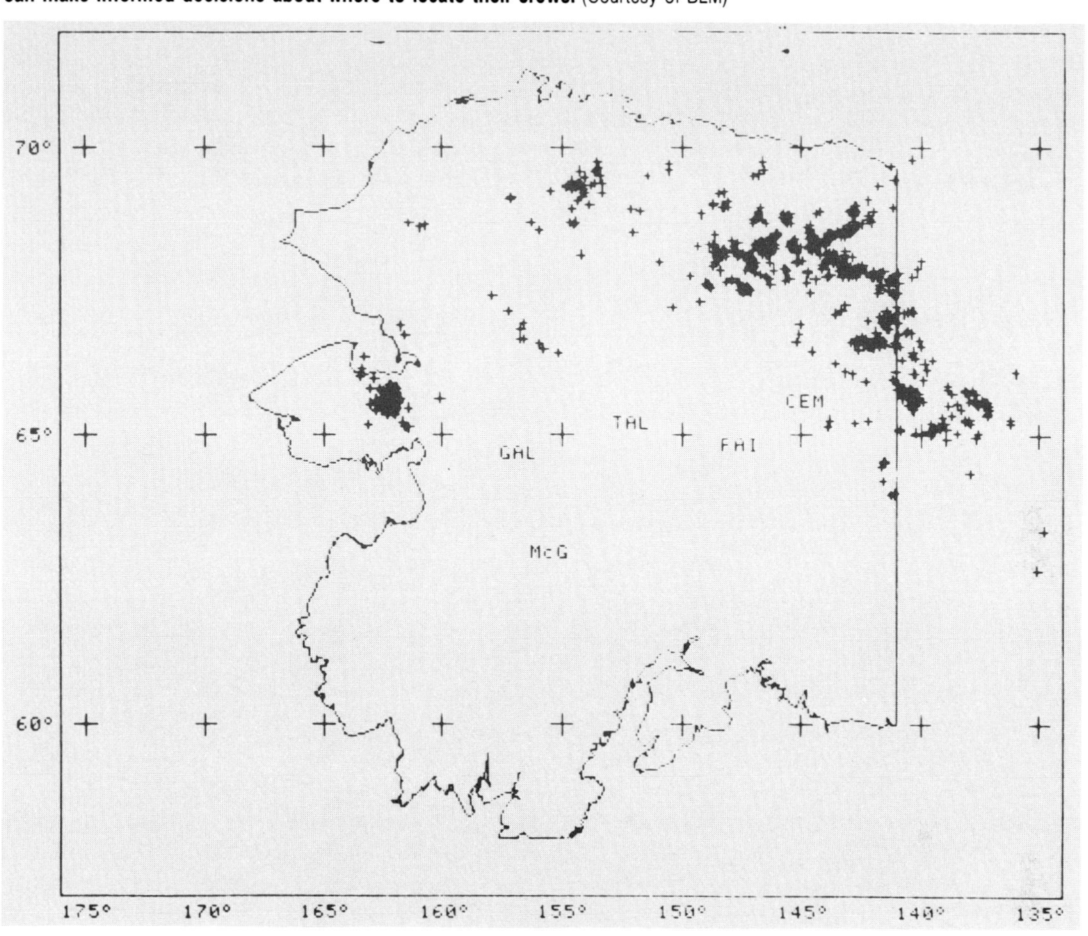

steady supply of such vegetation provided naturally by wildfire.

During critical winter months, caribou feed on lichen covering the taiga floor or clinging to the lower branches of trees. Wildfires destroy both trees and lichen; dry lichen on lower branches give fires an assist in climbing into the trees' crowns. Until the forest regenerates, burned areas offer little lichen for the caribou.

Fire may be necessary for long-term lichen production, however, since older forests often have declining lichen growth. In such stands, fire restarts the successional clock, and in 50 to 100 years, regenerated younger stands again produce bumper crops of lichen. Another bonus of recurring wildfire is that it helps establish the mosaic of different stands that serve as natural fire breaks, thus preventing monster fires from wiping out the entire forest: habitat, lichens and all.

Factors besides lichen affect caribou, including hunting, snow depth and pack, wolves, diseases, changes in migration patterns and probably others yet to be discovered. Caribou and wildfires have been in Alaska for centuries; we need to do more research to get the entire picture.

Smoke billows from a prescribed fire on the Kenai Peninsula. Such fires, set deliberately and carefully controlled, are used to improve wildlife habitat by destroying plants which compete with the animals' food. (John Warden)

Seven years after a fire in a spruce forest on the Porcupine River flats, willows have taken over. (Forestry Sciences Lab)

The site of a recent forest fire is blanketed with fireweed, so named because it is one of the first plants to grow over burned areas. (John Warden)

This aspen stand near Fairbanks, burned three months earlier by the big Rosie Creek fire, is revegetating quite rapidly. Aspen root suckers, which send up new sprouts, can be as dense as 100,000 per acre. (Walt Matell)

Wildfires can affect stream and river fish habitats by changing runoff from nearby forests. Early studies have shown little surface erosion from burned taiga and no temperature increase in streams after burning. But more research is needed.

Fire retardants dropped on fires may cause algae blooms in adjacent waterways if concentrations get high enough. Fire lines built to stop fires sometimes erode, causing siltation. This highlights the difficult decisions a fire boss must make: in the heat of the battle, for example, should a stream be jeopardized to save a valuable stand of timber?

Wildfires in the Coastal Forest

As might be expected, the rain-soaked coastal forests of Prince William Sound and southeastern Alaska do not burn very often. Fires there are usually caused by people, occur during exceptionally dry summers and are less than a few hundred acres. Exceptions include the 15,000-acre burn at Karta Bay on Prince of Wales Island at the turn of the century. Other large fires of the past have no written record, but have left large, **even-aged** second growth stands.

As in the Interior, forest managers in Tongass National Forest are considering using prescribed fires to improve wildlife habitat, and perhaps also to prepare sites for forest regeneration. They see fire as a natural part of the growth cycle of forests, and as a tool to be used.

Other Forest Resources

On a fundamental level, forests serve as vital processors and reservoirs of solar energy flowing through the planet's biosphere. When we use forest resources, we divert this flow for our benefit.

Timber and fuelwood are the two most obvious resources extracted from Alaska's forests, and the basis for a substantial industry, as we have seen. Forests are also playgrounds, for locals as well as people from around the world. They form backdrops for many of our communities; they protect soil stabiity and water quality. Forests moderate the weather and soothe the eye.

Forests are also home to fish and wildlife, a resource vital to subsistence hunters, trappers and fishermen, and the foundation of Alaska's fishing and guiding industries.

Wildlife

Many of Alaska's wildlife species can be found in forests, where they seek food and shelter. Some are permanent residents, while others spend only part of their lives in the woods.

Blue and spruce grouse are forest dwellers. During the summer these birds breed and forage on the forest floor and in nearby open areas, but in the winter they take to the trees, where they subsist on conifer needles. Canada geese nest and rear their broods in the forests of southeastern Alaska.

Snowshoe hares don't burrow. Instead, they spend their lives on the forest floor,

Black bears inhabit Alaska's forests, foraging and sometimes digging dens under the roots of trees.
(Wini Sidle)

A young bull moose rests in Denali National Park and Preserve. Moose eat aspen, birch and willow leaves and shoots. (Norma Dudiak)

traveling along established runs and seeking shelter under branches and logs. These animals eat twigs, roots, leaves, grass and bark. Hare populations fluctuate wildly: during peaks, as many as 600 hares per square mile have been recorded. Carnivores like lynx, wolves, wolverines, hawks, golden eagles and people prey on the hare.

Another forest resident, the noisy red squirrel, is well-equipped to move about in the trees, where it often has its nest. Constantly active during summer daylight hours, squirrels collect huge caches of spruce cones for winter food. They play an important ecological role by distributing tree seeds and fungus spores throughout the forest.

Furbearers also use forest habitats. Otters spend much of their time feeding in fresh or salt water, but use forest habitats as well, often digging dens under tree roots in nearby forests. Marten and mink are also found in many of Alaska's forests.

Porcupines eat the **cambium** layer of trees, gnawing under the bark, especially in the winter when more succulent vegetation is unavailable.

Beaver, of course, are highly dependent on trees. They eat the inner bark of trees, and log nearby forests for material with which to build their dams and lodges. Beaver usually choose trees between 4 and 12 inches in diameter, but chewed stumps up to 5 feet across have been recorded.

Larger mammals have ranges that include

▲A spruce grouse, common resident of the coastal forest, nestles in the understory near Homer. Grouse breed and forage on the forest floor during the summer. (Norma Dudiak)

►A snowshoe hare feeds on the bark of a fallen limb near Glennallen. (Don Cornelius)

A flock of snow buntings resemble white leaves on an otherwise bare cottonwood in the Matanuska Valley. (Don Cornelius)

▲**This tree was recently downed by a beaver near Ship Creek in southcentral Alaska. Beavers log forests for material with which to build their dams, usually selecting trees four to twelve inches in diameter.** (Lisa Holzapfel)

▼**Salmonberries and blueberries tempt summer hikers in the coastal forests.** (Paul Beck)

A worker from the Forestry Sciences Lab estimates understory plant cover and production. Such measurements are used to determine the value of a forest as wildlife habitat. (Forestry Sciences Lab)

forests. Moose eat apsen, birch and willow leaves and shoots during the early stages of forest succession, when the trees are still small. Bear will forage in forests, and some dig dens under tree roots or in logs. During the winter, mountain goats and deer descend to lower-elevation forests.

Resource Conflicts

Human development in forests can disturb wildlife, either directly or by altering habitats. Roads necessary for mining, timber harvesting or subdivisions, for example, increase "people-pressure" by making formerly remote areas accessible. Development often changes the frequency of fire, throwing natural ecological cycles out of balance.

Animals dependent on old growth forest habitat are adversely affected when such forests are logged and replaced with second growth. On the other hand, populations of animals that thrive in early successional stages of forest development — such as songbirds — can expand greatly. Hasty or

The Case of the Sitka Black-Tailed Deer

Biologists recognize that suitable habitat is essential for wildlife survival. One animal which has been receiving a lot of attention recently because of its habitat needs is the Sitka black-tailed deer.

This small deer, the most northerly of black-tailed deer, inhabits much of southeastern Alaska. On the mainland and larger islands, most of the population spends the summer above tree line, feeding on abundant and nutritious alpine and subalpine vegetation. As winter approaches and snow begins to accumulate, the deer descend to lower elevations. During harsh winters, if they are unable to find food under deep snow, many may die. Surviving deer congregate in old growth stands, where the sturdy canopy of large trees intercepts snow. Scattered openings in the canopy, created by fallen trees, let in the light needed by forage plants like bunchberry, goldthread and five-leaved bramble that grow in patches on the forest floor. Biologists call such areas "critical winter range."

Field studies to date have been conducted during relatively mild winters. One of these years, scientists should be able to learn how Sitka black-tailed deer react during a really severe winter with heavy snowfall throughout the Panhandle.

Until recently, clear-cut logging was thought to be beneficial for deer, because new clear-cuts quickly green-up with an abundance of light-loving forage plants. In southeastern Alaska, however, snowfall accumulates more rapidly in cutover areas than in adjacent uncut stands. About 25 years after harvesting, the second growth trees grow crowns large and dense enough to block most light from the forest floor. Forage plants disappear, and do not return until the area is cut again or until the stand evolves into old growth in 150 to 250 years.

The Forest Service is experimenting with ways to convert second growth stands into critical winter range for deer and other animals. Proposed treatments include thinning second growth in patterns that let in light to maintain some of the forage plants, and techniques for reducing debris left by logging. Such treatments will probably be expensive, and results are several years away.

Sitka black-tailed deer, residents of southeastern Alaska, descend to the shelter of coastal forests during the colder months of the year. (Paul Beck)

Much of southeastern Alaska's old growth forest is set aside in wilderness areas. Planned timber harvest on about 2.5 million acres of remaining national forest and private lands will convert old growth stands to second growth. If initial deer studies are correct, and unless the experimental second growth treatments work, deer winter ranges will be reduced accordingly. This loss is one of the costs of harvesting higher-volume timber stands in southeastern Alaska. Conversely, keeping deer winter ranges will result in an economic loss from harvesting lower-volume stands instead, or from reducing the timber harvest.

Blowdown into streams often prompts cleanup work to remove obstacles to fish passage. In the past, such cleanup was sometimes excessive, removing logs and debris necessary for shade, cover and streambank stability. (Mike Pease, USFS)

poorly planned developments can lower water quality of nearby streams, affecting fish and other animals miles away downstream.

Fish Live Here, Too

The effect of logging on salmon habitat has been a long-standing concern in Alaska. In 1909, the U.S. Forest Service and Bureau of Fisheries issued regulations prohibiting logging along streams with salmon hatcheries, and restricting logging near other salmon streams during summer and fall runs.

In the 1950s, when large-scale pulp logging started in southeastern Alaska, the Forest Service monitored adjacent salmon streams. They noted an increase in water temperature, more debris in the streams, and slight changes in stream sediment. The number of returning pink and chum salmon remained about the same after logging. At the same time, however, fish traps were abolished in Alaska; the increased number of fish thus able to return to fresh water may have masked possible decreases due to logging.

Resource managers agree that careless development can degrade fish habitat, and have developed ways to prevent such damage. In the past, roads sometimes

High winds can sometimes unravel forests standing next to logging units. Blowdown, such as this on Prince of Wales Island, can be prevented by designing units to minimize edges exposed to wind. (Wini Sidle)

Forest Service fisheries technician Bess Clark checks a stream for salmon fry using an electro-shocker, a device which lures fish into the scoop without harming them. (Walt Matell, USFS)

A biologist checks the amount of forest canopy above a fish spawning stream by viewing reflections in a level mirror. The canopy protects the stream and helps to maintain a fairly constant water temperature. (K. Koski, National Marine Fisheries Service)

disrupted the flow of streams; a well-designed road today will include culverts that allow fish to pass. Sediment-laden runoff from rock pits, landings, settling ponds and roads can be channeled away from fish streams. Logs can be yarded away from or lifted over streams, rather than dragged through them, as was once common practice.

If the forest is logged right up to the edge of a stream, the stream's ecology will change. For example, water temperatures fluctuate more than when the stream was protected under the forest canopy. Trees can be left in buffer strips along streams to prevent this, but winds often blow them over into the streambed. At one time, blowdown into streams was removed at great expense. Biologists now recognize that overzealous cleanup can be bad for the salmon, too, since a certain amount of logs and debris is actually necessary to provide them with resting pools, shade and nutrients.

Recently, scientists with the National Marine Fisheries Service have compared exposed streams in clear-cuts, streams with buffer strips and comparable streams in uncut old growth stands throughout southeastern Alaska. They found an array of similarities and differences, including more algae, more food organisms and a higher density of newly emerged coho fry in logged streams during the summer. On the other hand, they also found fewer year-old coho (soon to become the smolt that migrate to the ocean) in

Bald eagle nest trees are protected by law, and activities in their vicinity are controlled so the birds are not disturbed. (Walt Matell)

streams through clear-cut areas than in other habitats. Research is continuing to determine the relationship between logging and coho smolt production, and the impact of log dumps and raft storage on crabs and clams. With such data in hand, forest managers can design logging operations accordingly.

Protecting Habitats

State and federal governments have regulations designed to protect fish and wildlife habitat in Alaska. The Forest Service, for example, is required by law to ensure perpetuation of "viable populations" of all wildlife species found in national forests. Federal law restricts development within 330 feet of bald eagle nest trees, because the national bird depends on tall waterfront trees in which to nest and perch. The state's Forest Resources and Practices Act requires that logging not damage fish spawning streams in all forests in Alaska, including those on private lands.

We have much more to learn about forests and wildlife. In the long run, it appears that animals fare best in conditions under which they evolved. These conditions cannot always be preserved if human beings use a particular forest area for other resources. We can minimize habitat loss, and can even try to enlarge or simulate certain habitats. But we should also be candid and recognize that sometimes wildlife will be taking a back seat to other forest resources.

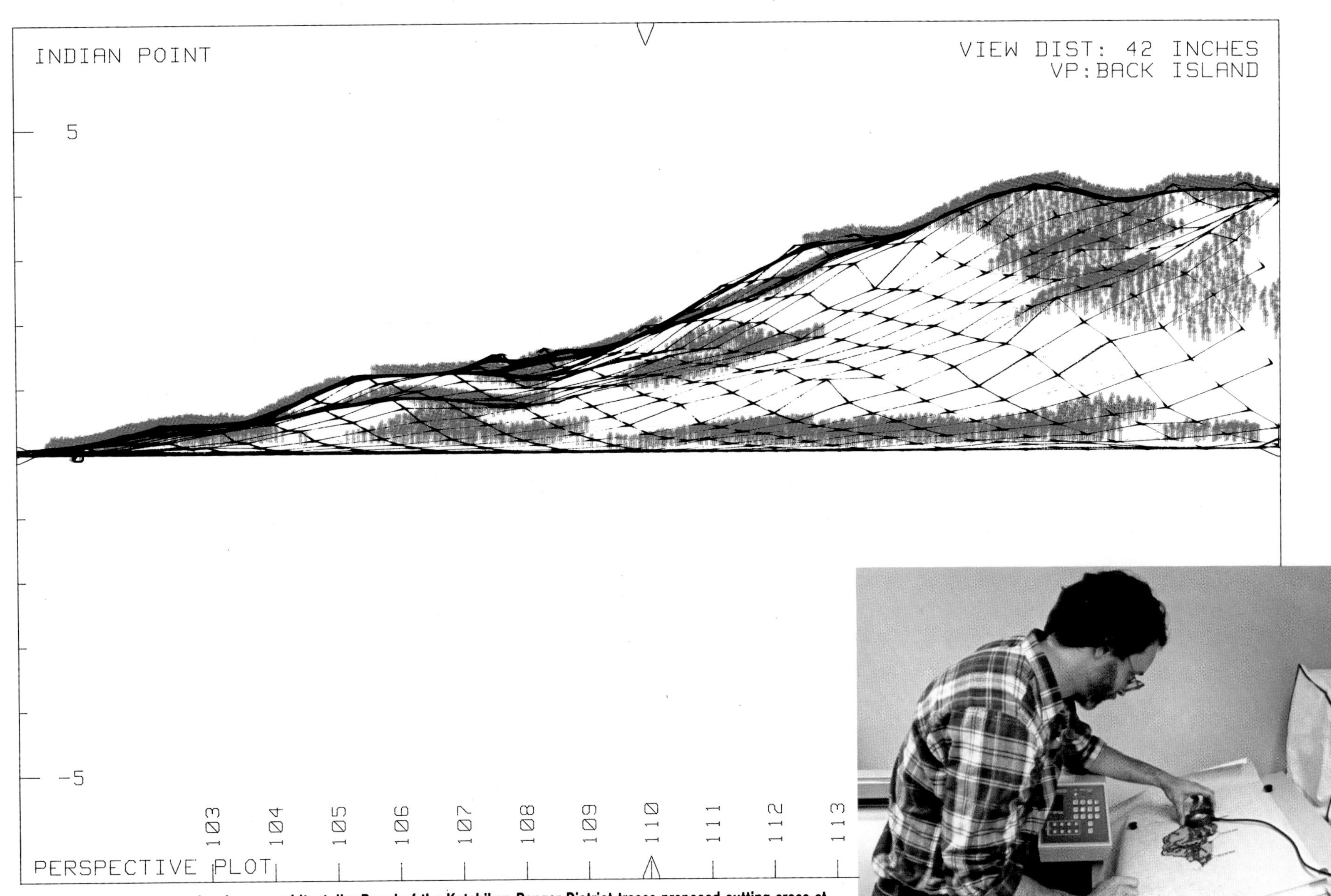

Landscape architect Jim Beard of the Ketchikan Ranger District traces proposed cutting areas at Indian Point on a table that senses pointer movements and converts them into digital information a computer can understand. The computer then combines this information with a stored digital model of the area's topography to produce printouts such as the one above, showing what the harvested area would look like from any specified viewpoint. (Right, Walt Matell; above, Courtesy of USFS)

Indian Point, near Ketchikan, seen here from nearby Back Island, was included in a plan proposed by the Alaska Department of Fish and Game and U.S. Forest Service which would have harvested some areas of Tongass National Forest heavily, while deferring other more sensitive areas. The plan was scrapped due to widespread public opposition. (Jim Beard, USFS)

Public Forests of Alaska

Outrageous waste and depredation of forests by "robber baron" industrialists in the late 1800s fanned popular sentiment for protecting the remaining public lands. Since most forestlands on the East Coast were by then privately owned, the focus shifted to the West — and to Alaska.

In 1891, Congress authorized establishment of "forest reserves" to protect natural resources on federal forestlands. The following year, President Benjamin Harrison, concerned that Pacific salmon might follow the decline of their Atlantic cousins, created the Afognak Forest and Fish Culture Reserve on that island north of Kodiak.

Eleven years later, President Theodore Roosevelt established the Alexander Archipelago Forest Reserve on the outer islands of southeastern Alaska. The reserve was enlarged to include the mainland in subsequent years, and the name was changed to Tongass National Forest. Chugach National Forest on the Kenai Peninsula and along Prince William Sound in southcentral Alaska was similarly created and enlarged in the early years of the century.

Twenty-five years after statehood, the state of Alaska had received much of its entitlement to more than 100 million acres. Alaska Natives are converting 44 million acres of selected public lands to private property under terms of the Alaska Native Claims Settlement Act of 1971.

Despite these changes, most of Alaska's

A camper has pitched his tent in the shelter of a cottonwood grove in Chugach State Park. People come from all over the world to enjoy the recreational opportunities in Alaska's forests. (Don Cornelius)

Hardy conifers hang onto sheer granite cliffs in Misty Fjords National Monument. (Linda Brownstein)

forests remain public. Let's take a closer look at some of them:

National Forests

Covering 16.9 million acres, Tongass is the largest unit in the National Forest system. It and the 5.8-million-acre Chugach are managed by the Forest Service, a branch of the U.S. Department of Agriculture, under a series of federal laws specifying multiple and sustained use of the forests' resources.

In these public forests, the Forest Service sells timber, maintains a network of recreation cabins, builds roads, manages fish and wildlife habitats, oversees mining, conducts forestry research, manages wilderness areas and ecological reserves, protects archaeological sites, and provides wildfire protection. The agency is also concerned about maintaining industries dependent on the forest's resources.

The Tongass timber sale program is unique among national forests because it includes two 50-year contracts [see "The Forest Industry Today," page 118]. Timber harvesting in the Tongass is at the core of a long-standing dispute between groups who want parts of the forest to remain undeveloped and supporters of industry who do not want to see large amounts of timber set aside.

In the late 1970s, the Forest Service developed a master plan for the Tongass. This land management plan subdivided the forest into small areas, each of which was to be managed for either wilderness, mostly roadless recreation, mixed uses, or mostly intensive timber harvest.

At the same time, Congress was considering legislation to establish parks, monuments, refuges and other conservation units on federal lands in Alaska. The Tongass was one focus of the debate, and a location for compromise. When passed, the Alaska National Interest Lands Conservation Act (ANILCA) of 1980 set aside 5.4 million acres of the Tongass as wilderness, allowed two mines to be developed in Admiralty Island and Misty Fjords national monuments (areas within the Tongass set aside for their ecological and aesthetic values), and mandated a timber harvest of 4.5 billion board feet per decade.

Only six percent of Chugach National Forest is considered productive forestland, so the harvest there has been low. Sitka spruce and western hemlock are the main commercial species, but stands of black cottonwood, Alaska-cedar and mountain hemlock can also be found. In the recently completed Chugach Forest Plan, the Forest Service proposes a 17-million board foot annual harvest.

National Forests are also home for wildlife and fish. Forest service biologists are discovering what needs they have, how logging affects habitat and how habitats can be improved. Determining seasons, bag limits, and other direct management of fish and wildlife populations in the Tongass and

Timber From the Tongass

Tongass National Forest is not wall-to-wall forest; ice fields and treeless areas constitute 43 percent of its acreage. Of the remaining forest, 5.7 million acres (one-third of the total acreage) is classified as commercial forestland. Of this land, 28 percent cannot be cut because it has been designated wilderness, and 37 percent is either considered unsuitable for timber harvesting or is retained for wildlife habitat or other uses. That leaves 35 percent (about two million acres) of the commercial forestland available for timber harvest.

Of the two million acres programmed for timber harvest, about 85 percent is old growth and 12 percent is newly established seedling/sapling stands, but little forest is **pole-timber** or young growth. This is exactly the pattern one would expect in the early decades of a mature forest being converted to second growth. Most of the seedling/sapling stands in the Tongass are very productive. This reflects recent harvest of old growth from these sites. Much of the best high-volume and accessible timber has already been cut.

Depending on timber markets, it may cost more to log the remaining lower-volume stands than the timber is worth. Therefore, to maintain the harvest at 4.5 billion board feet per decade, public investments are required.

The Forest Service is contracting precommercial thinning on second growth stands to improve future yield. Anticipating extra growth due to thinning of about 34 MMBF per year, they are allowing that much more timber to be harvested today. The government also plans to provide guaranteed loans to timber purchasers for advanced equipment such as long-reach yarders, helicopters and new systems to harvest timber in areas where existing equipment would not be economical or meet environmental standards. Roads into marginal stands will be built with permanent appropriations, rather than being financed by the value of the timber harvested.

In ANILCA, Congress established a $40 million per year Tongass Timber Supply Fund appropriation and a $5 million loan fund to pay for these programs.

Author Walt Matell stops to examine an alder during an outing in the coastal spruce-hemlock forest of southeastern Alaska. (Kathy Lucich)

A lone tree grows above the tundra in the Kotzebue "national forest," a purely whimsical designation in the Arctic. (Norma Dudiak)

Forest Service recreation cabins in remote areas of Tongass and Chugach national forests can be reserved for $10 per night. (Tom Buckhoe)

To avoid spongy, rain-soaked ground, hikers along the Naha River in Tongass National Forest use a boardwalk trail originally constructed by the Civilian Conservation Corps. (Walt Matell)

Chugach is the responsibility of the Alaska Department of Fish and Game. [See also "Other Forest Resources," page 168.]

Mining companies have a legal right to search for and extract minerals on national forests in areas that have not been specifically withdrawn from mineral entry. The Forest Service is charged with ensuring that mining causes minimal harm to the surface environment. When sufficient minerals are discovered, mining claim holders can apply to the Department of the Interior for a patent, which converts the claim to private property. Enclaves of such private forests are found throughout Tongass and Chugach.

People come from all over the world to enjoy the truly great outdoors of these two forests: they come to fish, hike, climb, hunt, kayak, photograph, pan for gold, canoe, study, beachcomb, ski or just kick back and relax in any of the 180 public cabins that rent for $10 per night. The Forest Service also maintains campgrounds, visitor centers and rustic airstrips. Aside from its intrinsic value, recreation — a resource the Forest Service measures in "visitor-days" — is a growing industry in Alaska.

Supporting local industries and communities, maintaining long-term productivity and diversity of the forest, and ensuring a balance of use and preservation are some of the lofty goals mandated in various laws that govern the national forests. By reviewing environmental impact statements issued on major national forest projects, the public can become involved in the planning that translates these goals into on-the-ground action.

State Forestlands

The State of Alaska now has title to about half of the commercial forestlands in the Interior, but only a small percentage of the coastal forest. Having just acquired these lands, and anticipating more conveyances, the Division of Forestry is still doing a lot of planning.

Driftwood below high tide line is considered property of the state, which issues permits for commercial salvaging. (Mary Ida Henrikson)

Alaska's constitution requires that "fish, forests, wildlife, grasslands, and all other replenishable resources belonging to the State shall be utilized, developed, and maintained on the **sustained yield** principle . . . " Like national forests, state forestlands must be managed so their future productivity will not be impaired. Aside from being a legal requirement, this also makes long-term economic and ecological sense.

The state legislature has endorsed plans for long-term management of forestlands. In 1983, it established the 1.6-million-acre Tanana Valley State Forest along the Tanana River east and west of Fairbanks. In the Chilkat Valley, northwest of Haines, 229,000-acre Haines State Forest Resource Management Area will be managed for multiple-use of its forest resources. The designation includes a reserve along the Chilkat River for the thousands of bald eagles that congregate there each winter.

The Chilkat Valley is also the location of an ongoing state timber sale to the mill at Haines. Started in 1979, the 15-year, 10.2 MMBF-per-year contract was specifically designed to provide employment in an area hard hit by the poor economy. As the timber market slump continued, the mill had financial difficulties, closed, and reopened under new management in 1984. The new owner, Pacific Forest Products, Inc., plans to mill some of the wood into lumber and truck it to the Interior.

The state has sold timber from other areas as well. During the last 25 years, about 900 MMBF of sawtimber, house logs and cordwood have been harvested from sales throughout the state. Two are notable: a salvage sale near Tyonek on upper Cook Inlet that turned 73 MMBF of beetle-killed white spruce into wood chips, and the just-completed 206 MMBF sale from Icy Cape on the narrow strip of coastal forest below Mount Saint Elias.

Other forestlands managed by the state are being considered for state parks or locations for future communities. Some forests have already been transferred to private ownership through the state's land disposal program.

Other Public Forests

Until recently, federal forestlands not included in national forests or parks remained under the jurisdiction of the old General Land Office, a bureau charged with disposing public domain lands to miners and homesteaders. It was mainly a custodial agency, and made few attempts to manage resources.

By mid-century, the newly created Bureau of Land Management inherited the public domain lands. The best forestlands had already been reserved by other agencies.

Beetle-killed white spruce logs are loaded onto a truck at the state's Tyonek salvage sale in 1976. During the 10 years the sale operated, 74 million board feet of damaged spruce were harvested.
(John Galea, USFS)

Karen Lewandoski and Dan Wieczorek of the state Division of Forestry in Fairbanks compare aerial photographs of the Rosie Creek area, which burned in 1983, with a photomosaic showing the layout of salvage timber sales for the area. The state is increasing active management of its forestlands.
(Walt Matell)

A pair of bicyclists braves rain and gravel roads to enjoy the beauty and solitude of this southeastern forest. (Walt Matell)

Today, more are being conveyed to native corporations or the State of Alaska. Therefore, although BLM still manages a lot of land in Alaska, little is considered productive forestland.

Like the U.S. Forest Service and the State Division of Forestry, BLM is also a multiple-use agency. It maintains watersheds, soils, wildlife habitats and campgrounds, and sells some timber from the public forests. Most sales are for fuelwood, but some small ones have been for sawtimber. BLM is responsible for fire protection on much of the Interior's forests and tundra.

Other federal forestlands in Alaska have been reserved from the public domain for various purposes throughout the years. In the national wildlife refuges, forests are managed primarily for habitat. Forests in national parks and preserves are allowed to evolve naturally, except that fires may be controlled in areas close to facilities or with prior-use rights.

We are indeed fortunate that a hundred years ago people with foresight started the tradition of public forests in this country. Whether working forests or forests left undisturbed, they are our common treasure. As stewards of these lands, we have the responsibility to ensure that they are managed well. What will our descendants a hundred years from today say about *our* foresight?

Hikers enjoy the fall colors near Thunderbird Creek in Chugach State Park. (Don Cornelius)

This aerial photo shows forest regeneration on Tuxekan Island in southeastern Alaska. The roaded area in the foreground was clear-cut one year before the photo was taken; the green growth along the shore beyond is taking place on the site of another clear-cut, harvested in the 1950s. (Walt Matell, USFS)

Glossary

BACKFIRE	To deliberately set a fire to burn away vegetation in front of an advancing wildfire, thus depriving it of fuel.
BIOMASS	In the context of using forests for energy production, the total amount of fiber grown by plants in a given area.
BOARD FOOT	A unit of measure equal to a board one inch thick and one foot square.
BROWSE	Young shoots and foliage used as food by deer, caribou, moose and other wildlife.
BTU	British thermal unit, a measure of the energy required to raise one pound of water one degree Fahrenheit. A 100-watt light bulb burning one hour consumes 34 Btu. A cord of paper birch contains about 21 million Btu.
BUTT-SWELL	The area at the base of a tree where it flares out into the root system. Because the grain is twisted, this part of the log is unusable for lumber.
CAMBIUM	Tissues under the bark responsible for a tree's diameter growth.
CANOPY	The collective crown (upper branches and leaves) of trees in a forest.
CANT	A log that has been sawed on two sides. Buyers run them through mills a second time to turn them into timbers or lumber.
CATKIN	A long cluster of many tiny flowers.
CLEAR-CUTTING	A harvest method that removes all trees from an area, called the cutting unit.
COMMERCIAL FORESTLAND	Land producing or capable of producing crops of industrial wood, economically accessible and not withdrawn for other uses. In coastal forests, it must have at least 8 MBF per acre; in the Interior, it must grow at least 20 cubic feet per acre per year.

COMMERCIAL THINNING — Thinning of trees large enough to have commercial value. See also **THINNING.**

CONK — Shelflike fruiting bodies of fungi, found on the outsides of rotting trees.

CROWN — Branches and leaves at the top of a tree. A wildfire is said to "crown" when it spreads from the ground into the treetops.

DBH — "Diameter at breast height," a convenient height standard for measuring the diameter of trees. "X-inch dbh" means that a tree has an X-inch diameter four and one-half feet from the ground.

DIOECIOUS — Plants having male and female flowers on different plants. **MONOECIOUS** plants, on the other hand, have both sexes on the same plant. Willows, poplars, cottonwoods and aspen, for example, are dioecious.

DOMINANT — The tallest trees in a stand.

DRAINAGE CUT — A large clear-cut that harvests most of the commercial trees in a valley at one time. Although the trend has been toward dispersed, smaller clear-cuts, some land managers prefer to cut some areas heavily, leaving others uncut.

ECOLOGICAL SUCCESSION — See **SUCCESSION.**

EVEN-AGED — A stand in which most trees are about the same age and size.

FIXING — Conversion, by bacteria on roots of legumes and trees such as alder, of atmospheric nitrogen into compounds necessary for plant growth.

FLUTING — Growth of tree trunks with buttresses, creating trunks with uneven cross sections. Fluting is considered a defect in hemlock.

GROWING STOCK VOLUME — The volume of live trees of commercial species, excluding seedlings, saplings and trees so defective that they couldn't be made into lumber.

KRUMMHOLZ — Gnarled and slow-growing trees, caused by adverse conditions like constant wind or poor nutrition. Literally, "crooked wood."

LANDING — Location, usually on the side or end of a road, to which logs are yarded, and where they are loaded onto trucks.

LAYERING — A means of reproduction in which the buried lower branches of trees sprout stems that become new trees.

LUMBER — Manufactured wooden boards, squared on all four sides.

MBF — Thousand board feet.

MMBF — Million board feet.

MUSKEG — Open, boggy area with waterlogged organic soils, too wet and acidic for most trees.

MYCORRHIZAE — Fungi that help plant roots absorb nutrients.

NET ANNUAL GROWTH — Increase or decrease in the volume of wood of commercial timber in a stand per year. When managed for timber, it is considered good practice to harvest a stand before its net annual growth declines.

NEW GROWTH — See **SECOND GROWTH.**

OLD GROWTH — Stands in which most dominant trees are past physiological maturity, considered more than 150 years old in southeastern Alaska. Such stands are characterized by mature trees, dying trees and younger trees replacing the dead ones.

OVERSTORY — Trees in a stand whose canopies rise above lower-growing vegetation.

PATCH-CUT — A series of small clear-cuts separated by uncut forest.

PEELER — A high-quality log suitable for making plywood. The log is put on a giant lathe, and literally peeled into a continuous sheet of thin veneer. Veneers are then sandwiched, glued and pressed into plywood.

PERMAFROST	Permanently frozen ground found in northern regions. The upper portion of permafrost, called the active layer, may thaw during the summer, but refreezes the following winter.
PHEROMONE	Hormones released by insects and animals to attract others of the same species.
POLE-TIMBER	Trees larger than saplings, but smaller than sawtimber.
PRECOMMERCIAL THINNING	Thinning of stands with trees too young to be of commercial value. The cost of such thinnings must be recovered by increased future yields.
PRESCRIBED FIRE	Deliberate setting of forest or shrub fires to achieve some desired end. Most often used to promote browse vegetation for wildlife.
REFUGIA	Areas that were not coverd by ice during times of glacial advance, and thus served as refuges for plant and animal life.
RELEASE	Rapid growth of remaining trees after thinning.
RHIZOME	Underground stem that turns upward to become a new plant.
ROTATION	In a forest managed for timber production, the time between planned harvests. Optimal rotation for fiber production is to cut the second growth before its growth rate slows.
SAMARA	The winged seed of trees like the maple.
SANITARY CUT	Cutting a stand to remove trees infected with disease or insects.
SAPLING	Tree between 1 and 5 inches in diameter.
SAWLOG	A log at least 8 feet long, with a minimum small-end diameter of 6 inches for softwoods and 8 inches for hardwoods, reasonably sound and straight, suitable for milling.
SAWTIMBER	Trees of commercial species suitable for harvest.

A logger cuts a dead tree, called "snag." Snag is considered a safety hazard in logging areas, and is therefore removed. (Caribou Trails Photography)

SECOND GROWTH	Stand of trees replacing one removed by harvest or act of nature. Commonly used as synonym for young growth or new growth.
SEED-TREE SYSTEM	Harvesting system similar to clear-cutting except some trees are left to distribute seeds over the cut.
SEEDLING	Tree less than one inch in diameter.
SELECTION SYSTEM	Harvesting system in which only some trees are removed in each of a series of cuttings. Aim of this method is to perpetuate an uneven-aged stand.
SELECTIVE CUTTING	Harvesting only good-quality trees; high-grading.
SHELTERWOOD	Harvesting system which leaves some trees to protect the new second growth. When the new stand is well-established, the remaining shelterwood is also harvested.
SILVICULTURE	The science of growing trees as crops for human benefit.
SITE PREPARATION	The silvicultural practice of creating optimal conditions for re-establishment of new stands of trees following harvest.
SNAG	A dead yet still standing tree, valuable for wildlife habitat but hazardous to loggers working in the woods.
STRINGER	A stand of trees along a watercourse.
STUMPAGE	The price paid by a purchaser for standing timber.
SUCCESSION	The order in which plant species colonize a barren site, or re-establish themselves on a disturbed site. Also called ecological succession.
SUSTAINED YIELD	The policy of cutting timber on a given land base at a rate no faster than the forests can be expected to grow back in the future. Mandated by Congress for national forests by the Multiple-Use Sustained Yield Act of 1960, and by Alaska's constitution for state lands.
TAIGA	The evergreen forests of the high latitudes, including those of interior Alaska. From the Russian word meaning "land of little sticks."
THINNING	**Cutting trees in a stand to decrease its density, thereby improving growing conditions for the remaining trees.**
TREE	A woody plant having an erect trunk at least 3 inches in diameter, a crown of foliage, and growing to at least 12 feet when mature.
TREE LINE	The dividing line between areas where trees grow and where they don't. Tree line is encountered going up tall mountain slopes, and also occurs geographically. (See map on page 15.)
TUNDRA	Treeless, rolling or flat terrain beyond tree line.
UNDERSTORY	Vegetation on the forest's floor or under the forest's canopy.
UNEVEN-AGED	A stand in which the trees are of different ages and sizes. Old growth stands are usually uneven-aged.
VEGETATIVE REPRODUCTION	Propagation by non-seed means, such as rhizomes, sprouting from stumps or layering.
YARDING	Pulling logs out of the woods.
YOUNG GROWTH	An immature stand of trees. See also **SECOND GROWTH.**

Metric Conversions

one hectare	=	2.47 acres
one acre	=	0.405 hectares
one cubic meter	=	35.3 cubic feet
one cubic foot	=	0.0283 cubic meters
one cord	=	2.5 to 3.0 cubic meters (approximate)
one MBF	=	4.5 cubic meters (approximate)

Alaska Geographic® Back Issues

The North Slope, Vol. 1, No. 1. The charter issue of *ALASKA GEOGRAPHIC®* took a long, hard look at the North Slope and the then-new petroleum development at "the top of the world." *Out of print.*

One Man's Wilderness, Vol. 1, No. 2. The story of a dream shared by many, fulfilled by a few; a man goes into the Bush, builds a cabin and shares his incredible wilderness experience. Color photos. 116 pages, $9.95.

Admiralty . . . Island in Contention, Vol. 1, No. 3. An intimate and multifaceted view of Admiralty: its geological and historical past, its present-day geography, wildlife and sparse human population. Color photos. 78 pages, $5.00

Fisheries of the North Pacific: History, Species, Gear & Processes, Vol. 1, No. 4. The title says it all. This volume is out of print, but the book, from which it was excerpted, is available in a revised, expanded large-format volume. 424 pages. $24.95.

The Alaska-Yukon Wild Flowers Guide, Vol. 2, No. 1. First Northland flower book with both large, color photos and detailed drawings of every species described. Features 160 species, common and scientific names and growing height. Vertical-format book edition now available. 218 pages, $12.95.

Richard Harrington's Yukon, Vol. 2, No. 2. The Canadian province with the colorful past *and* present. *Out of print.*

Prince William Sound, Vol. 2, No. 3. This volume explores the people and resources of the Sound. *Out of print.*

Yakutat: The Turbulent Crescent, Vol. 2, No. 4. History, geography, people — and the impact of the coming of the oil industry. *Out of print.*

Glacier Bay: Old Ice, New Land, Vol. 3, No. 1. The expansive wilderness of southeastern Alaska's Glacier Bay National Monument (recently proclaimed a national park and preserve) unfolds in crisp text and color photographs. Records the flora and fauna of the area, its natural history, with hike and cruise information, plus a large-scale color map. 132 pages, $11.95.

The Land: Eye of the Storm, Vol. 3, No. 2. The future of one of the earth's biggest pieces of real estate! *This volume is out of print*, but the latest on the Alaska lands controversy is detailed completely in Volume 8, Number 4.

Richard Harrington's Antarctic, Vol. 3, No. 3. The Canadian photojournalist guides readers through remote and little understood regions of the Antarctic and Subantarctic. More than 200 color photos and a large fold-out map. 104 pages, $8.95

The Silver Years of the Alaska Canned Salmon Industry: An Album of Historical Photos, Vol. 3, No. 4. The grand and glorious past of the Alaska canned salmon industry. *Out of print.*

Alaska's Volcanoes: Northern Link in the Ring of Fire, Vol. 4, No. 1. Scientific overview supplemented with eyewitness accounts of Alaska's historic volcano eruptions. Includes color and black-and-white photos and a schematic description of the effects of plate movement upon volcanic activity. 88 pages. *Temporarily out of print.*

The Brooks Range: Environmental Watershed, Vol. 4, No. 2. An impressive work on a truly impressive piece of Alaska — The Brooks Range. *Out of print.*

Kodiak: Island of Change, Vol. 4, No. 3. Russians, wildlife, logging and even petroleum . . . an island where change is one of the few constants. *Out of print.*

Wilderness Proposals: Which Way for Alaska's Lands? Vol. 4, No. 4. This volume gives yet another detailed analysis of the many Alaska lands questions. *Out of print.*

Cook Inlet Country, Vol. 5, No. 1. Our first comprehensive look at the area. A visual tour of the region — its communities, big and small, and its countryside. Begins at the southern tip of the Kenai Peninsula, circles Turnagain Arm and Knik Arm for a close-up view of Anchorage, and visits the Matanuska and Susitna valleys and the wild, west side of the inlet. *Out of print.*

Southeast: Alaska's Panhandle, Vol. 5, No. 2. Explores southeastern Alaska's maze of fjords and islands, mossy forests and glacier-draped mountains — from Dixon Entrance to Icy Bay, including all of the state's fabled Inside Passage. Along the way are profiles of every town, together with a look at the region's history, economy, people, attractions and future. Includes large fold-out map and seven area maps. 192 pages, $12.95.

Bristol Bay Basin, Vol. 5, No. 3. Explores the land and the people of the region known to many as the commercial salmon-fishing capital of Alaska. Illustrated with contemporary color and historic black-and-white photos. Includes a large fold-out map of the region. *Out of print.*

Alaska Whales and Whaling, Vol. 5, No. 4. The wonders of whales in Alaska — their life cycles, travels and travails — are examined, with an authoritative history of commercial and subsistence whaling in the North. Includes a fold-out poster of 14 major whale species in Alaska in perspective, color photos and illustrations, with historical photos and line drawings. 144 pages, $12.95.

Yukon-Kuskokwim Delta, Vol. 6, No. 1. This volume explores the people and life-styles of one of the most remote areas of the 49th state. *Out of print.*

The Aurora Borealis, Vol. 6, No. 2. Here one of the world's leading experts — Dr. S.-I. Akasofu of the University of Alaska — explains in an easily understood manner, aided by many diagrams and spectacular color and black-and-white photos, what causes the aurora, how it works, how and why scientists are studying it today and its implications for our future. 96 pages, $7.95.

Alaska's Native People, Vol. 6, No. 3. In this edition the editors examine the varied worlds of the Inupiat Eskimo, Yup'ik Eskimo, Athabascan, Aleut, Tlingit, Haida and Tsimshian. Included are sensitive, informative articles by Native writers, plus a large, four-color map detailing the Native villages and defining the language areas. 304 pages, $24.95.

The Stikine, Vol. 6, No. 4. River route to three Canadian gold strikes in the 1800s. This edition explores 400 miles of Stikine wilderness, recounts the river's paddle-wheel past and looks into the future. Illustrated with contemporary color photos and historic black-and-white; includes a large fold-out map. 96 pages, $9.95.

Alaska's Great Interior, Vol. 7, No. 1. Alaska's rich Interior country, west from the Alaska-Yukon Territory border and including the huge drainage between the Alaska Range and the Brooks Range, is covered thoroughly. Included are the region's people, communities, history, economy, wilderness areas and wildlife. Illustrated with contemporary color and black-and-white photos. Includes a large fold-out map. 128 pages, $9.95.

A Photographic Geography of Alaska, Vol. 7, No. 2. An overview of the entire state — a visual tour through the six regions of Alaska: Southeast, Southcentral/Gulf Coast, Alaska Peninsula and Aleutians, Bering Sea Coast, Arctic and Interior. Plus a handy appendix of valuable information — "Facts About Alaska." Approximately 160 color and black-and-white photos and 35 maps. 192 pages. Revised in 1983. $15.95.

The Aleutians, Vol. 7, No. 3. Home of the Aleut, a tremendous wildlife spectacle, a major World War II battleground and now the heart of a thriving new commercial fishing industry. Contemporary color and black-and-white photographs, and a large fold-out map. 224 pages, $14.95.

Klondike Lost: A Decade of Photographs by Kinsey & Kinsey, Vol. 7, No. 4. An album of rare photographs and all-new text about the lost Klondike boomtown of Grand Forks, second in size only to Dawson during the gold rush. Introduction by noted historian Pierre Berton: 138 pages, area maps and more than 100 historical photos, most never before published. $12.95.

Wrangell-Saint Elias, Vol. 8, No. 1. Mountains, including the continent's second- and fourth-highest peaks, dominate this international wilderness that sweeps from the Wrangell Mountains in Alaska to the southern Saint Elias range in Canada. Illustrated with contemporary color and historical black-and-white photographs. Includes a large fold-out map. 144 pages, $9.95.

Alaska Mammals, Vol. 8, No. 2. From tiny ground squirrels to the powerful polar bear, and from the tundra hare to the magnificent whales inhabiting Alaska's waters, this volume includes 80 species of mammals found in Alaska. Included are beautiful color photographs and personal accounts of wildlife encounters. 184 pages, $12.95.

The Kotzebue Basin, Vol. 8, No. 3. Examines northwestern Alaska's thriving trading area of Kotzebue Sound and the Kobuk and Noatak river basins. Contemporary color and historical black-and-white photographs. 184 pages, $12.95.

Alaska National Interest Lands, Vol. 8, No. 4. Following passage of the bill formalizing Alaska's national interest land selections (d-2 lands), longtime Alaskans Celia Hunter and Ginny Wood review each selection, outlining location, size, access, and briefly describing the region's special attractions. Illustrated with contemporary color photographs. 242 pages, $14.95.

Alaska's Glaciers, Vol. 9, No. 1. Examines in-depth the massive rivers of ice, their composition, exploration, present-day distribution and scientific significance. Illustrated with many contemporary color and historical black-and-white photos, the text includes separate discussions of more than a dozen glacial regions. 144 pages, $10.95.

Sitka and Its Ocean/Island World, Vol. 9, No. 2. From the elegant capital of Russian America to a beautiful but modern port, Sitka, on Baranof Island, has become a commercial and cultural center for southeastern Alaska. Pat Roppel, longtime Southeast resident and expert on the region's history, examines in detail the past and present of Sitka, Baranof Island, and neighboring Chichagof Island. Illustrated with contemporary color and historical black-and-white photographs. 128 pages, $9.95.

Islands of the Seals: The Pribilofs, Vol. 9, No. 3. Great herds of northern fur seals drew Russians and Aleuts to these remote Bering Sea islands where they founded permanent communities and established a unique international commerce. Illustrated with contemporary color and historical black-and-white photographs. 128 pages, $9.95.

Alaska's Oil/Gas & Minerals Industry, Vol. 9, No. 4. Experts detail the geological processes and resulting mineral and fossil fuel resources that are now in the forefront of Alaska's economy. Illustrated with historical black-and-white and contemporary color photographs. 216 pages, $12.95.

Adventure Roads North: The Story of the Alaska Highway and Other Roads in *The MILEPOST*®, Vol. 10, No. 1. From Alaska's first highway — the Richardson — to the famous Alaska Highway, first overland route to the 49th state, text and photos provide a history of Alaska's roads and take a mile-by-mile look at the country they cross. 224 pages, $14.95.

ANCHORAGE and the Cook Inlet Basin . . . Alaska's Commercial Heartland, Vol. 10, No. 2. An update of what's going on in "Anchorage country" . . . the Kenai, the Susitna Valley, and Matanuska. Heavily illustrated in color and including three illustrated maps . . . one an uproarious artist's forecast of "Anchorage 2035." 168 pages, $14.95.

Alaska's Salmon Fisheries, Vol. 10, No. 3. The work of *ALASKA*® magazine outdoors editor Jim Rearden, this issue takes a comprehensive look at Alaska's most valuable commercial fishery. Through text and photos, readers will learn about the five species of salmon caught in Alaska, different types of fishing gear and how each works, and will take a district-by-district tour of salmon fisheries throughout the state. 128 pages, $12.95.

Koyukuk Country, Vol. 10, No. 4. This issue explores the vast drainage of the Koyukuk River, third largest in Alaska. Text and photos provide information on the land and offer insights into the life-style of the people who live and have lived along the Koyukuk. 152 pages, $14.95.

Nome: City of the Golden Beaches, Vol. 11, No. 1. The colorful history of Alaska's most famous gold rush town has never been told like this before. With a text written by Terrence Cole, and illustrated with hundreds of rare black-and-white photos, the book traces the story of Nome from the crazy days of the 1900 gold rush. 184 pages, $14.95.

Alaska's Farms and Gardens, Vol. 11, No. 2. An overview of the past, present, and future of agriculture in Alaska, and a wealth of information on how to grow your own fruit and vegetables in the north. 144 pages, $12.95.

Chilkat River Valley, Vol. 11, No. 3. Its strategic location at the head of the Inside Passage has long made the Chilkat Valley a corridor between the coast and Interior. This issue explores the mountain-rimmed valley, its natural resources, and those hardy residents who make their home along the Chilkat. 112 pages, $12.95.

ALASKA STEAM, A Pictorial History of the Alaska Steamship Company, Vol. 11, No. 4. An inspiring story by Northwest history writer, Lucile McDonald, of men and ships who pioneered the hazardous waters of the northern travel lanes to serve the people of Alaska. Over 100 black-and-white historical photographs. 160 pages, $12.95.

The Northwest Territories, Vol. 12, No. 1. This issue takes an in-depth look at Canada's immense Northwest Territories, which comprise some of the most beautiful and isolated land in North America. Supervising editor Richard Harrington has brought together informative text and color photos covering such topics as geology and mineral resources, prehistoric people, native art, and the search for the Northwest Passage. Also included is a look under the ice of the Canadian Arctic. 136 pages, $12.95.

NEXT ISSUE:
Alaska Native Arts and Crafts, Vol. 12, No. 3. This issue takes an in-depth look at the art and artifacts of Alaska's native people. Author Susan Fair presents information on native lifestyles through the years, discussing the evolution of ancient tools and ceremonial items into contemporary works of art. Old and new artwork is illustrated in more than 200 full color photos. Also included is a chapter on archaeology in Alaska written by Robert Shaw. To members in August 1985. Price to be announced.

All prices U.S. funds.

The Alaska Geographic Society

Box 4-EEE, Anchorage, AK 99509

Membership in The Alaska Geographic Society is $30 (U.S. funds), and includes the following year's four quarterlies which explore a wide variety of subjects in the Northland, each issue an adventure in great photos, maps, and excellent research. Members receive their quarterlies as part of the membership fee at considerable savings over the prices which nonmembers must pay for individual book editions.